New Science, New World

Denise Albanese

NEW SCIENCE, NEW WORLD

Duke University Press   Durham and London

1996

Dedicated to the memory of

Terry A. Comito and Anthony Albanese

for lines to live by:

"I once more smell the dew and rain" and

"There are three men in a rowboat

and the oars leak . . ."

Contents

Acknowledgments

It occurs to me that writing acknowledgments is not really about putting paid to a debt. Signs of gratitude don't discharge obligations, but instead model them as dischargeable—which doesn't seem very likely to be true. Rather too like the act of writing the book to which they are prefixed, these acknowledgments make me hopelessly aware of how much will be left unsaid, how poorly the little that does get said represents the personal and intellectual terrain of obligation.

But here goes, anyway. A book as long in the making as this one has accrued a sedimentary history—which in turn means many successive layers of indebtedness to be worked through, brought to light, made evidentiary to a kind of historicist consciousness that is, I guess, not very far from the thematic interests of this book. Graduate school, for instance: although little of *New Science, New World* resembles the long-ago dissertation on Bacon he directed, John Bender has pride of place for his early enthusiasm about the possibility of conjoining my physics past with my literature-based present. I am also indebted to the friends from those days, Thomas Moser Jr., Deborah Laycock, and Deidre Lynch, and to the Whiting Foundation, for the award through Stanford University of a Whiting Dissertation fellowship in 1984–1985, which greatly aided my initial researches.

Subsequent ties to people and institutions—George Mason University, the Folger and Huntington Libraries, Wesleyan University, and elsewhere—have sustained me in many ways both concrete and intangible. On the first count: The former chair of the English department, Johannes Bergmann, arranged unexpected and generous institutional support at a critical moment. I am grateful to the Folger Library for a

Acknowledgments

Grant-in-Aid for the Spring of 1988; to the Huntington Library for a research fellowship for the summer of 1989, and to Wesleyan University, whose Mellon Fellowship at the Center for the Humanities in 1989–1990 gave me the opportunity to recast the project in drastic and necessary ways. The staff of the Folger Library, like the staff of the Huntington, more than live up to their reputations for efficiency and solicitude. And so, for that matter, did the staff at the Center for the Humanities.

Other colleagues and friends helped in less material, but no less important, ways; among them Steve Brown, Dina Copelman, Sheila ffolliott, Werner Gundersheimer, Lorna Irvine, Deborah Kaplan, Rosemary Jann, Barbara Melosh, Eileen Sypher, Ellen Todd, and Steven Weinberger. This book would not be what it is without the benefit of many conversations and contestations, at the Folger, at Wesleyan, and in Washington, D.C. Nancy Armstrong, Catherine Belsey, Karen Bock, Michael Bristol, Dympna Callaghan, Ed Casey, Christina Crosby, Ann Cvetkovich, Inderpal Grewal, Kim Hall, Caren Kaplan, Lindsay Kaplan, Tom Moylan, Áine O'Brien, Richard Ohmann, Eric Smoodin, Gary Spear, Leonard Tennenhouse, Valerie Traub, and Alok Yadav, all engaged with my work in one form or another—reading drafts or responding to lectures, soliciting papers and essays, challenging my premises and augmenting my bibliography. In so doing, they tested it (and me) in ways that can only have improved the book, if not the temperament of its author. No one could imagine a better—wittier, more humane, more intellectually sympathetic—senior colleague than Devon Hodges. Zofia Burr's insistence on the prosaic doability of intellectual work was almost as valuable to me as her keen readings of some pretty muddled prose. And David M. Levy offered patience, information, computer support, and the occasional ideological cuddle. I wish every writer an editor like Ken Wissoker. Pam Morrison and the production staff at Duke University Press have expended every effort to make *New Science, New World* a beautiful book; any errors are of course my responsibility.

Finally, I wish to thank Metro Pictures for permission to reproduce

Acknowledgments

Cindy Sherman's *History Portraits;* the Folger Library, for the engravings by de Bry and van der Straet; and the Library of Congress, for the illustrated title page from Vesalius. I am grateful to the editors of *English Literary History* for permission to use my article, *"The New Atlantis* and the Uses of Utopia," *ELH* 57 (1990): 503–528.

Too many people close to me have died during the writing of this book. I will never be able to replace Gary Spear in my life, and there are not enough words to say why. The others—my father, Anthony Albanese, and my former teacher and colleague, Terry A. Comito—had the grace to see me through almost to the end of this project. *New Science, New World* is dedicated to these two men, who were so formative, so necessary, and so difficult, to live with and to lose. Hail and farewell.

PORTA tenet primas; habeas, GERMANE, secundas; / Sunt, GALILAEE, tuus tertia regna labor. / Sidera sed quantum Terris caelestia distant, / Ante alios tantum Tu, GALILAEE, nites. / Hi TELESCOPIO metantur paucula passum / Millia telluris, vel vada salsa freti; / Quos infinitis, clarum dum scandis Olympum / Arte parato OCULO, passibus ipse praeis. / Cedas, VESPUCI, cedatque COLUMBUS; uterque / Ignotum saltem per mare tentat iter; / Nec plane antipodum tellus tamen inscia priscis, / Nec quondam astronomos fugit uterque polus: / Sed tu stellarum seriem, nova sydera caeli, / Humano generi qui daret, unus eras.

> —Poem to Galileo by Johannes Faber, prefixed to *Il Saggiatore*

Another error . . . is a distrust that any thing should now be found out, which the world should have missed and passed over so long time, as if the same objection were to be made to time that Lucian maketh to Jupiter and other the heathen gods, of which he wondereth that they begot so many children in old time and begot none in his time. . . . So it seemeth men doubt lest time is become past children and generation, wherein contrariwise we see commonly the levity and unconstancy of men's judgments, which, till a matter be done, wonder that it can be done, and as soon as it is done, wonder again that it was no sooner done. . . . And the same happened to Columbus in the western navigation. But in intellectual matters it is much more common. . . .

> —Francis Bacon, *The Advancement of Learning,* Book I

And surely it would be disgraceful if, while the regions of the material globe—that is, of the earth, of the sea, and of the stars—have been in our times laid widely open and revealed, the intellectual globe should remain shut up within the narrow limits of old discoveries.

> —Bacon, *The New Organon,* LXXXIV

any differential system is sustained by that which also fails to sustain it; were that not the case, there would be no history.

> —Jonathan Goldberg, *Writing Matter*

Introduction

As the first three epigraphs to this Introduction suggest, the "New World" and the "New Science," isomorphic for us through a trick of language that retrospectively endows each with an equal title to novelty, were also seen as homologous in the seventeenth century. What, however, can the homology be read to signify from this vantage? To claim both Europe's geographical expansionism and the transformation of knowledge-seeking agendas as "modern" is a usual enough historical description. Still, it leaves unaddressed the near-circularity of the formulation and hence the genealogical relationship such "modern" forms have with the time and space of the writing that describes them.

New Science, New World reads the isomorphism of novelty as a *symptom* of modernity—as the sign of changes beneath the skin of early modern culture, changes that reveal how scientific modernity emerges from within the humanist textuality of the late Renaissance. Although it concerns some of the many explicit links made in seventeenth-century texts between the "New World" and the "New Science," my book is not a study of how the connections are represented, how the organized investigation of nature is domesticated—propagated—through a rhetoric of common novelty. Nor is it a systematic account of such connections, although a very useful study remains to be written on just that subject. Rather, I propose to consider these colonialist tropes as the discursive signs of cultural change in suspension—what Raymond Williams has usefully called a structure of feeling.

To demonstrate how the literary becomes the exotic other of the scientific—in other words, to show how science and literature have come to occupy opposite poles in (post)modern culture—my book seeks out the connections between the New World and the New Sci-

ence as simultaneously emerging discursive patterns. While the repeated joining of the two topics in Renaissance texts makes clear that a
rhetorical analogy exists between colonialism and science, the following analysis makes more of the conjuncture. Through examining utopian structures within texts canonical either in literature or science,
I argue that such structures provide a way to read the onset of taxonomies of writing and, in turn, of a cultural division of labor, an
inaugural scripting of an opposition between "fact" and "fiction" that
follows along other oppositional constructions of culture still dominant in postmodern culture.

The present study takes its inspiration from Michel de Certeau's
assertion in *The Writing of History* that the modern world produces itself
through othering, through discursive and material mechanisms that
effectively bifurcate regions of culture, the better to legitimate some
and delegitimate others.[1] The first, and in some ways the most significant, instance of this differentiation occurs when the present is separated from the past and what had heretofore been a lived archive becomes a repository of texts. But, as de Certeau argues, this rupture
inaugurates something like a simultaneous chain reaction in which
nature and "labor," and then discourse and the body, begin to be deployed oppositionally. (It may be said in passing that this hypothesized
moment of rupture constitutes the starting point for most of Foucault's
genealogical projects.)

In beginning with a shift in the discourse of temporality, *The Writing of History* makes the problematic of causation in historiography a
discursive event in itself. It is humanist historiography that recognizes "change," that makes time linear and progressive, that creates the
"need" for cause-and-effect narratives—and not the other way round.
For this reason alone, de Certeau's text is invaluable for scholars who
wish to question the necessarily reductive models of causation in historical narration, but who also find the successive descriptions of horizons limiting in the questions they make possible. But *The Writing of
History* has the virtue of casting historical isomorphisms as the mechanisms of cultural change, in addition to their obvious role as the signs

of such change. Hence, the pertinence of the terms "New World" and the "New Science" in this study. I argue that, as a consequence of the culturally productive of mechanisms of opposition—particularly those mobilized by Renaissance colonialism—the emergence of modern scientific ideology in the seventeenth century resulted in the positing of fiction, of literary representation, as its binary (and prospectively devalued) opposite. One might consider fetishism as an intermediate term, in part because etymologically it straddles the epistemological gap between fiction and fact.[2] Although the networks connecting fetishism, fact, and fiction are not a direct part of the subject to hand, it is crucial to note that fetishism gains utility as a discursive counter at the emergence of colonialism. This, in turn, suggests the more general significance of the "New World" as a site of differentiation and distinction between "true" and "false"—or fictional. The legendary stories of dogheaded men, Amazons, and cannibals that attach themselves to the newly discovered territory are gradually replaced by accounts of New World inhabitants that, as I argue in chapter 1, are nascently akin to ethnography.

The dichotomy between fact and fiction, like all ideological handles, is of course subject to contestation, as too schematic and hence too neat. The cultural theorist Donna Haraway, for instance, begins her influential *Primate Visions* by questioning what she calls the "moral obligat[ion] to oppose fact and fiction."[3] Haraway, trained in biology as she is, embraces no reductive dismissal of scientific practice—indeed, as her work on "situated knowledges" suggests, she is interested in how to recover the study of nature from its institutionalized transgressions.[4] Nevertheless, since scientific rationality has historically constituted a regime of truth that has subjugated other epistemological systems as its "others," Haraway deems it crucial to call into question the current lines of opposition along which difference (and disciplines) are still constructed, and power deployed.

But what if we inverted the order of reasoning? What if, instead of problematizing the *present* discursive authority of science by showing how its facts are (among other things) productive and interested artifacts, we hypothesized the historical emergence of a difference between "fact" and "fiction"? What might the conditions be for such a differentiation to occur—not in any absolute sense, but in the local sense made available through a retrospective reading of some crucial seventeenth-century texts? To a great extent, raising such questions is the project of this book. The word "hypothesize" above is not casually chosen. As chapter 1 indicates, one sign of an emergent attempt to distinguish the scientific from the poetic, or fictional, is the controversy over the status of hypotheses as representations. Indeed, Fernand Hallyn has equated the early modern hypothesis with the practice of poetics.[5] As just as Hallyn's analyses are, their formalistic bent suppresses the prior question: why is it necessary to suggest that science has a "poetics" at all? Why, that is, work to establish not just a connection, but a near-identity between these discourses so clearly distinguished within the modern faculty of disciplines in terms of power, prestige, and epistemological authority?

Hallyn's study, like mine, entails a form of redress. Like him, I invoke the possibility of a historical a priori different from the classical one out of which "science" proper formally emerges. But discerning the phantasms of preemergence, whether of science or of "literature," is fraught with particular difficulties in the case of the early modern period, which the choice of the word "poetics" does not escape. For one thing, the earlier seventeenth century lacks the formal institutions that serve as material undergirding for the mature disciplines that go under the signs I have sometimes used. Hence the importance of figuration: tropes and discourses of novelty, particularly those associated with colonialism, stand in for those institutions. At the same time, they offer a way to make connections across disparate practices, and to take seriously the constitutive function of language.

A brief overview of this study makes clear just how far-flung those practices have become. Chapter 1, "Making It New: History and Novelty in Early Modern Culture," constitutes a species of prolegomenon. It seems necessary to begin any discussion of the "New Science," however revisionary, by an examination of the place of novelty in late Renaissance culture. The way into this discussion may seem unexpected, since it commences with some illustrations of "History" produced by the artist Cindy Sherman. These consist of uncanny attempts both to inhabit and reproduce "the past," even as they emblematize the impossibility of doing so seamlessly. This oblique discussion serves to introduce topics—gender, the body, novelty, and the past—that will continue to circulate throughout the book. To further understand the staging novelty in these discussions of new worlds and new sciences in early modernity, I then read Donne's *Ignatius His Conclave.* As a text uniting Columbus and Copernicus as damnable agents of the new, *Ignatius His Conclave* makes it possible to read rupture and innovation as functions of an emergent universalism of temporality, itself a function of modernity. In contrast to humanism's suppression of the gap separating its present from the classical antiquity which authorizes it (witness Petrarch's letters to classical authors), New World modernity provides an alternative way of talking about the past. I then move to a critical, exemplary analysis of the illustrations to Thomas Harriot's *Briefe and True Report of the New Found Land of Virginia.* Such (familiar) colonialist texts as Harriot's frame the radical difference between the inhabitants of Europe and the Americas as evidence of a universal narrative of development. If classical humanism posited the past as father to the present, the discovery of "primitive" cultures in the Americas suggests an alternative view—of the past as infantile, and of the present, consequently, as advanced. (That the past gestates, develops into the present, is not itself a universal model; rather, it competes with an elegiac representation of bygone wholeness or fecundity in late Renaissance texts.) In the framework provided by New World modernity, the bodies of natives, of others, become material evidence; this implicit dichotomy between corporeality and disincorporation maps onto the emergent scientific dichotomy between subject and object, which becomes paradoxically visible under the sign of femininity.

Chapter 2, "Admiring Miranda and Enslaving Nature," attends to the sign of that femininity, which is to say to the work that gender does as a category of historical analysis (to borrow Joan Scott's useful phrase). I cast some of Donna Haraway's insights in *Primate Visions* about the ideological vectors of race and gender in modern science back into the late Renaissance to establish the pertinence of her analytical categories for early modern literary texts—themselves deemed "literary" by a back-formation that counterposes the producing of fiction with the inscribing of fact. The central text is *The Tempest,* which has been much read as a reworking of materials about colonialism by interpreters of the English Renaissance. I do not intend to contest that now-dominant reading so much as to supplement it with an older one: Prospero as magus/"New Scientist." It is no accident that Shakespeare's romance has lent itself to analyses of colonialist structure and scientific theme: New World and New Science are here contiguous, and the structures of domination and othering in the play-text place the work of the mind over against the work of the body, as Prospero is placed over Caliban. Just as excavations of the colonialism implicit in *The Tempest* operate to bracket Miranda as a character, so the triumphant suppression of the corporeal which the play makes possible has woman as ground, as pretext, as the state of nature upon which ideologies of modern science (and modern subjectivity) are constructed. The textual embodiment of the New World, in turn, is the discursive counterpart of that state of nature: the condition of possibility for the emergence of "science."

Utopian form as synecdoche for literary humanism, and as index for a moment of cultural instability, is further explored in chapter 3, "*The New Atlantis* and the Uses of Utopia." To read Bacon is to be poised on the threshold separating the literary canon from the scientific one: on one side, Milton and Shakespeare dwell, and the other is inhabited by Galileo. The liminality of Baconian texts is a clear trace of the cultural differentiation that the book is concerned to examine, and it is mapped out by *The New Atlantis*'s own relation to More's *Utopia,* its colonial pretext. Fiction is used to valorize fact, and the New World becomes a textual mechanism for the production of scientific subjects. Yet the incompleteness of Bacon's text is the triumph of the residuum, of the unclear line demarcating Renaissance from modern, of the humanist text from scientific propaganda.

Chapter 4, "The Prosthetic Milton; Or, The Telescope and the Humanist Corpus," extends the previous argument about the encroachment of scientific modernity into literary space through a reading of *Paradise Lost.* A text embodying both the dominant forms of the Renaissance and the nascent structures of modernity, the epic's staging of human inquiry becomes symptomatic of an emergent ideology of knowledge. In constructing its own universal truths about Man [*sic*] and the cosmos out of the materials of theology and literary humanism, Milton's epic attempts to (re)establish the humanist text as an alternative to scientific models, and Eden as an America immune to the temporal narratives of modernity. But, as the central conversations between Raphael and Adam indicate (especially that on heliocentrism), the free play of knowledge is a sign of that modernity that cannot be excluded, and the American paradise an unstable utopia that cannot but slip over into history and contingency, and to the coming of another system of universals.

The final chapter, "Galileo, 'Literature,' and the Generation of Scientific Universals," opens up the issues I have considered within English print culture of the seventeenth century by returning to the Italian framework within which I commenced my examination of humanism. Here, I examine Galileo's exemplary construction of experimental spaces for an emergent scientific modernity. While the various documents Galileo sends out along with his scientific treatises invoke aesthetic criteria to justify—or protect—the performance of astronomy, in effect these citations of the literary also open up the possibility for reading a nascent, productive difference between one type of text and another. Given the ideological volatility of the Copernican hypothesis in the seventeenth century, Galileo's request for a literary response to, rather than a strict evaluation of, its truth-claims, cannot be seen as sign of an Edenic time before discursive differentiation, a time only of "writing." Instead, it signifies a betrayal that such differentiation has (always?) already occurred. Then I turn to Galileo's *Dialogue Concerning the Two Chief World Systems:* the (textual) space of indifference that Bacon invents for inquiry, through his assimilation of narrative forms to empirical study, may be taken into the purview of the thought-experiment, which I read as a utopian text, a linguistic palimpsest of the New World, much as it is in Bacon. As with other representations of the New World I have considered, the thought-experiment models a space beyond constraint—beyond materiality and corporeality, indeed, in the freedom that it provides to examine interdicted ideas, a space beyond Catholic ideology: a claim itself constitutive of the ideology of modern science.

A few words about my method and critical practice seem in order. In a study dedicated, in part, to hypothesizing the emergences of the rationalistic and evidentiary structures generally associated with scientific modernity, I have often chosen, not systematic demonstration and analysis, but something more "literary." This is, of course, not to enlist on the side of belles lettres, nor to suggest a bias against theory: my indebtedness to models of ideology critique, discourse analysis, historicisms "new" and after, to Haraway, Althusser, Foucault, de Certeau, and Serres is everywhere apparent. But I also wanted to practice a rather different type of historical modeling: as my epigraph from Jonathan Goldberg (himself inspired by Derrida) evinces, this modeling depends on a differential practice that does not reach its destination, or, rather, that survives in (post)modernity as an oblique form of critical practice. Perhaps the best methodological gloss on the pages that follow can be found in Foucault's offhand speculations in *The Order of Things:*

In the modern age, literature is that which compensates for (and not that which confirms) the signifying function of language. Through literature, the being of language shines once more on the frontiers of Western culture—and at its center—for it is what has been most foreign to that culture since the sixteenth century; but it has also, since this same century, been at the very center of what Western culture has overlain. (44)

I say more in chapter 2 about Foucault's nostalgic positioning of literature as a conduit to authentic raw being; certainly despite himself Foucault also reveals the less innocent and more material role literature has played as a form of colonial indoctrination "on the frontiers of Western culture." The subjugating function of literature is very much to the point. But so, I might add, is its potential status as a subjugated discourse. Not, of course, that literature is not (still) mythologized as a site of "truth": but the utility of those quotation marks makes clear the nature of the problem. In invoking "literature" for my practice, I aim to valorize, not a model of transcendent universalism, but an almost Renaissance notion of a regime of truth with but an indirect relationship to the factual. The decision not to follow a systematic analytic also accounts for the uneven attention accorded to some strains of the argument, such as gender. While questions around embodied binaries are prominent in the chapters on Shakespeare, Milton, and historiography, they appear but intermittently in other sections. Since I am concerned to excavate a series of parallel oppositions whose formation (or re-formation) occurs more or less simultaneously, it makes sense that in certain texts certain aspects of that project might be foregrounded and others occluded.

I shall say more about my use of the word "literature" later. Before that, I want to address the status of Foucault in *New Science, New World*. To some, Foucault may seem an odd name to adduce to any account of sciences other than exclusively the human-centered. Such human sciences stand in a more-or-less mimetic relation to the authorizing mode of the natural sciences, with their "hard" data, mathematical methods, and valorization of empiricism. But these methods converge on an object of study that is self-evidently refractory, unlike the exteriorized nature recorded by apparatuses of investigation like the telescope. And even more: they are deployed around and oriented toward an object of study that was the concern of official interests and culture, so that the science of man could be read critically as at once establishing claims to power-knowledge and serving the interests of the state in the management of individuals or populations.

These quasigovernmental issues seem far away from the pure talk of the heavens or the subatomic, or (to pose the question of distance another way) the at-times inchoate systematizing of natural phenomena with which I am concerned. And yet it must surely be true that the production of "man" as object of knowledge is in dynamic, if subsequent, relationship to a "nature" similarly opened to scrutiny, brought into being as a discursive object, and rendered knowable in relation to the privileged subject-position from which the gaze, in the Foucauldian sense, emerges. In fact, the very secondariness of Foucault's institutions—medicine, psychology and psychiatry, and the like—reveals his assumptions about the legitimacy (in terms of power-knowledge) of the prior discourses of an objectified nature.

In an introduction to Georges Canguilhem's *The Normal and the Pathological,* Foucault has forged an explicit connection between the historical researches of French scholars like Canguilhem, and the Frankfurt School—whose well-known members, Max Horkheimer and Theodor Adorno, were also most famously critical of scientific modernity.[6] Because it so clearly lays out the tradition from which Foucault's own work emerges, the introduction is worth quoting at length:

Works such as those of Koyré, Bachelard, or Canguilhem could indeed have had as their centers of reference precise, "regional," chronologically well-defined domains in the history of science but they have functioned as important centers of philosophical elaboration to the extent that, under different facets, they set into play this question of the Enlightenment which is essential to contemporary philosophy.

If we were to look outside of France for something corresponding . . . it is undoubtedly in the Frankfurt School that we would find it. . . . [I]n the end both pose the same kinds of questions, even if here they are haunted by the memory of Descartes, there by the ghost of Luther. These questionings are those which must be addressed to a rationality which makes universal claims while developing in contingency; which asserts its unity and yet proceeds only by means of partial modification when not by general recastings; which authenticates itself through its own sovereignty but which in its history is not perhaps dissociated from inertias, weights which coerce it, subjugate it. In the history of science in France as in German critical theory, what we are to examine essentially is a reason whose autonomy of structures carries with itself the history of dogmatisms and despotisms—a reason which, consequently, has the effect of emancipation only on the condition that it succeeds in freeing itself of itself.[7]

Foucault's figuring of this kinship leads, finally, to a note on the terms I employ. While I have generally bypassed historiographic contestation over the extent to which it is possible to speak of a "Copernican revolution," I realize that I have stepped willingly into another quagmire in writing of "the Renaissance," "humanism," "modernity," and especially "literature" and "science." My debt to Foucault is everywhere apparent, and I did not lightly forgo the strenuously won insights his texts provide about the incommensurability of successive discursive regimes, nor the need to question traditional periodicity. Indeed, in working to establish a resemblance between two dominant discourses with equal title to novelty, I have had to bear in mind the importance of horizontal over vertical relationship, of filiations in space rather than over time. In *The Archaeology of Knowledge,* Foucault emphasizes that the classical discourse of "Natural History" differed, in procedures and in the organizational strategies around statements, from a "comparable" discourse of flora and fauna in the sixteenth century: so it may be said that the "natural philosophy" of the seventeenth century is far from identical with the culturally weighted discourse of modern institutional science.[8]

The differential specificity of a formation or an object of study is worth preserving. Yet it has seemed more urgent to supplement the archaeological Foucault with the genealogical one, and, indeed, to move toward the more overtly political critiques of modernity's dominant epistemologies afforded by Marxist models (which may have an explicit relation to Foucault), and feminist analyses (which often have had to contest the gender blindness of Foucault's schemata of power). Anachronistic terms have genuine strategic utility here, since they have the virtue of keeping the argument clearly focused on the present-day stakes of historicist argument. Moreover, they can be qualified, contested, reinscribed, deployed against the apparently teleological and therefore reductive tendencies of a naive engagement with questions of historical continuity and difference. Written large in the immensely useful passage I have cited is precisely the sort of philosophical abstraction seldom associated with Foucault. "Reason" is a way to figure forth the dominant ideology of what I have throughout termed "modernity," and such abstractions, when understood historically, seem to me worth retaining.

Perhaps the best comment on my practice is provided by Cindy Sherman's "History Portraits," a consideration of which inaugurates the study that follows. As these photographs suggest in all their energetic and grotesque perversity, history in the time of postmodernity has begun to betray our investment in it, has begun to let that long-silent corpse speak, snigger, mock our official accounts. The past may indeed be prologue—but it may also, and inevitably, be a parody of the present. As I will go on to argue, forgery functions as the mirror image of humanist history, which is to say historicism as it has generally been practiced. As if through a distorted glass, forgery shows us the hopes and desires of historians who, since the Renaissance emergence of the discipline, have wanted access to the past "as it really was," to provide an authoritative because truly representative rendering of bygone affairs, events, persons, texts.

Sherman's pictures are anything but such forgeries, and so are anything but historiography played straight. Again and again, they raise the issue of originals and imitations, of true and false, only to dismiss them as ludicrous, as deceptive manifestations of a will to assimilate the past, mistaken both in its predication and its pretense of universal knowledge. What the "History Portraits" offer instead is a clamorous reappropriation of the canonical past, wholly suitable for critically engaged work in early modern cultural studies. In showing how that past was scripted, they also show how it can be scripted differently.

Making It New: History and Novelty in

Early Modern Culture

A text and two sets of images provide an entry into the argument
that follows. The first set is a series of photographs taken by the
artist Cindy Sherman and generally referred as the "History Portraits."[1]
The second set is probably more familiar to readers of Renaissance culture, since it comes from Theodor de Bry's much-discussed engraved
illustrations to Thomas Harriot's *Discovery and Report of the New Found
Land of Virginia* (1590). And the text is John Donne's *Ignatius His
Conclave* (1611), a moralized exploration of the place of novelty, and
hence of modern cultural formations, at the end of the Renaissance in
the seventeenth century. Each point of entry, whether text or image,
makes a problem out of power and knowledge as they function in the
telling of history; in so doing, each maps out the complex interrelationships between past and present, European and North American, and
mind and body as constituents of modernity.

As these symptomatic readings will indicate, historical narratives
are predicated equally on imagined relations and tactical silences; they
demand an ideological adjudication between what may be compre-
hended as familiar and what must be suppressed, or investigated, as
alien. This is true in all three instances, whether those narratives flag
the emergence of the early modern subject of humanism, the "new"
modernity betokened by colonialism and science in the seventeenth
century, or, as in the case of Sherman, the postmodern representation of
identity and estrangement on which the previous two meet.

In fact, the argument begins out of chronological order; it traces
and so completes a backward trajectory, and in so doing it studies the
problematic of historical retrospection. As meditations, belated ones,
on the modern project, Sherman's images make visible categories of

representation, technology, and the body produced by gender that are only teasingly, evanescently apprehensible in the early modern text and illustrations. In this regard, Sherman's photographs are also easy to read. To be sure, they are complex as images; but they emerge out of a recent critical terrain that has conditioned self-consciousness about apparatuses of social and aesthetic reproduction, and the range of subject positions that emerge from such reproduction. With the necessary correctives, this self-consciousness can usefully be cast backward. Nevertheless, because the "History Portraits" are something of a departure for Sherman, they have not much been read as I propose to do—as forays into historicity, as skeptical interventions into the presentation and production of historical consciousness.

Sherman's previous work—primarily the "Untitled Movie Stills" of the late 1970s—exploited and problematized the position of Woman as object of the Lacanian gaze in a series of portraits that evoked the style and narratives of the genre movies of the 1940s and 1950s. In each of the earlier photographs the disguised and costumed Sherman is caught by the camera she herself sets up to catch her, situated in some enigmatic yet hauntingly familiar narrative space. Whether reaching for a book, overtaken while walking on dark city streets, or surprised as she ducks out a sliding glass door in a full-length slip, Sherman inhabits the clothing, demeanor, and desirability of stereotypical femininity as produced by U.S. movie culture.

The twist is of course obvious, and it is this twist that has given this earlier work its totemic status as feminist critique.[2] In her formal control over the circumstances of image production, Sherman both induces a voyeuristic response and challenges one's right to it. Since she can be located on both sides of the camera, her assumption of the pose of movie woman makes clear that femininity thus defined as object of the gaze is a performance. Further, the tantalizing incompleteness of the narrative space she set up makes something else clear as well—that a lot more is involved here than meets the eye, at least as framed by the classic fetishistic technology of the cinema.

But the neat dovetailing between Sherman's work and a feminist theory that owes much to Lacan and film practice comes to grief in the "History Portraits," photographs that play off of the burnish—both formal and ideological—of Old Master paintings. In these portraits Sherman is no longer posing as the seductive and alluring woman: not one of these is a Venus, or any other allegorical female in fetching dishabille. Frequently, in fact, her encounter with (art) history does not leave her a woman at all. She is as likely to be a burgher as a Madonna, and even quotes Caravaggio's Bacchus in one of the few portraits to set her up as a focus of erotic desire, however overdetermined (figure 1). Moreover, the "History Portraits" make full, even ludicrous use of the prostheses that have otherwise been the subject of her camera (as, for example, in the "Untitled" series of 1984). Fake body parts abound, from breasts (perhaps not surprisingly) to noses to artificially elongated foreheads (figures 2, 3). The resulting photographs seem at first merely to invite laughter. In fact, so incisive a critic as Laura Mulvey suggests that these images "lack the inexorability and complexity of her previous phase": she reads them primarily as efforts to draw attention "to the art-historical fetishization of great works and their value."[3]

Of course, such a critique constitutes part of the photographs' agenda: how else to account for the attempt to stage—and to fall short of—the rich patina of oil in the glossy and superficial intransigence of the photograph? But the reason I begin with these photographs lies precisely in their break with the overtly "feminist," and all but decontextualized, career of the photographic image. If the "History Portraits" are not "inexorable" (a position that oddly privileges a linear clairvoyance), it is perhaps because they do not constitute a continued engagement with the problematics of the modern sex-gender system as defined solely within contemporary theories of gender and the gaze. Rather, these uncanny pictures offer a way to begin theorizing the historical beyond the limitations of canonical postmodern usage.[4]

Fredric Jameson has provided an influential account of the period construct that has come to be called postmodernity. He places the subject of late capitalism in what is effectively a house of mirrors: "the postmodern" relates to the historical past primarily as a crucial absence,

1. Cindy Sherman, Untitled (History Portrait # 224, 1990; by permission of Metro Pictures, New York).

2. Cindy Sherman, Untitled (History Portrait # 225, 1990; by permission of Metro Pictures, New York).

3. Cindy Sherman, Untitled (History Portrait # 211, 1990; by permission of Metro Pictures, New York).

a loss which can be charted through, for example, architecture that knows and summons history only through a deracinated and eclectic citation of period style. If seen through this particular grid, Sherman's portraits are merely confirmations of that loss, reproductions of a pastness that cannot be inhabited and understood as the basis for meaningful political action, but only parodied.

I would not be the first to note that Jameson's model—or, for that matter, Jean Baudrillard's more anarchic version of the flight of significance—of postmodernity is totalizing, and so crucially blind to issues of power and representation attendant on nonhegemonic subjectivities. The evacuation of meaning from history that he laments is far from total, and far from disabling. On the contrary, when history's master narratives of dominance and opposition (as Jameson recognizes them) lose their cogency, heretofore marginalized subjects are afforded a space in which to produce an alternative critical understanding of past formations.

This space is where Sherman's photography operates. In contrast to Jamesonian despair and the loss of authenticity, or Baudrillardian glee and the play of surfaces, Sherman's forgeries of the past announce their false provenance and so offer a critique of dominant art-historical narratives of subjectivity, embodiment, and gender. What they fake is the nonmodern body and its habitus; in calling too much attention to the materiality of the signifier, her portraits, whose subject is "history," cannot but inflect the signified.

As a result, the history which Sherman is interested in staging is the imagined history of embodiment, which means that these representations are politicized, like her earlier work. But the "History Portraits" implicate the subject in ways other than the film simulacra had called forth. Sherman's earlier images, with all their suaveness, tended perhaps to duplicate the very ideal object they also critiqued, or at least to be too readily misread as mere duplicates.[5] From the apparently truth-producing glamour of film, Sherman has moved to the recovery of mundane particularity, even ugliness, suppressed by the idealizing and aestheticizing work of Old Master painting. Thus Norman Bryson's discussion of the "abject" as the inevitable corollary of Sherman's project in these stills—the return of something very like the repressed, the unruly flesh that bulges all too abruptly in the photographs from the confines of the proper and painterly geometric swells we recollect in their debasement. For him, Sherman's "colonization" and subsequent jettisoning of Old Masterism demonstrates "the violence of the ego's insertion into social personae that clamp down over the flesh like a carapace or a prosthesis or an iron mask."[6]

Bryson's reversion to the language of psychoanalysis, to an ego punished and policed by form, does much to connect the "History Portraits" to the film stills, and even to the later works that focus on objects of disgust. And yet by circumscribing the "self" within the portraits, it offers effect for explanation and renders the bounded ego a captive rather than a product of such representational practices. In contrast, Sherman's practice questions the priority of interior life. More overtly than ever before, her truck is with the embodied subject, per-haps even with the body "itself"—but not merely as observed or repres-sive (or repulsive), but resolutely as constructed, as material historical artifact. Hence the rhetorical efficacy of prostheses, which remind the viewer that different cultural formations produce the apparent truth of the flesh. Hence, too—and this is worth underscoring—the gender variability of her subjects. The freedom Sherman affects in inhabiting male as well as female subjects suggests that the photographs work to make a problem of the apparently immutable sex-gender system of modernity that gave her earlier work its point. Apparently immutable because apparently factual: and here the question of medium and tech-nology, both as deployed and as foregrounded, comes into play.

Oil paint works as part of an apparatus of "artistic" representation, and it more obviously constructs its subject than the cold lens of the camera claims to do. Although the camera's claim to objectivity is readily deconstructed, the photographic lens is the latest development in a technology whose history is successive reproductions of "the real." It seems useful, therefore, to stress the function of the lens within the ideology of scientific objectivity, especially in the lens's medicalized form as an instrument for inquiry into the deep, and so presumably determining, structures of the human body.[7] In a sense, the lens thus conceived competes with an older mode of aesthetic, epistemological, and social reproduction, the portrait in oils.

Although it seems perverse to juxtapose these two moments in the history of social reproduction on the one hand, and of the production of images on the other, that is in effect what Sherman has done. Her (self-)portraits, especially those imitating Renaissance models, dramatize moments of discursive coalescence. But they also put those moments into question. Here, they seem at once to assert and deny, is staged the "birth of the individual"; here, too, is staged the "art" that stresses the "exceptional subject," either canonized by religious discourse (all those Madonnas) or by bourgeois wealth (the brocaded profiles, for example).

But notice, too, that I suggested the "History Portraits" problematize these moments of formation; this also gets back to my suspensions of their status as self-portraits, a subject I will have more to say about. Here, again, the prostheses enter in: their arrant artificiality, preposterous breasts on Madonnas, bizarrely unlikely faces, remind us not just of the artifact that is the body in history, but of the symbolic violence that lies submerged just beneath the humanist call to identify our subjectivities with those depicted in "great art." In a sense, that act of interpellation effectively disembodies the subject who gazes—a tacit dematerialization that Sherman's inhabitations correct through their insistence on the priority of the body as signifier. Only through the grotesque self-distortion her prostheses represent, she seems to say, can we locate ourselves then and there.

Note that I used the word "seems": the question is whether there is any other way to get there, to read the past, to narrate relation without the collapse into identity. Sherman suggests the gap from here to there is not smoothly negotiated, that we bring along ourselves and our interpretive agendas, much as her camera intrudes on bodies meant only for oil. But I disagree with Mulvey in thinking these representations are avowals only of a body fetishism consequent upon loss, a position to which Mulvey's theoretical engagement with Freud and Lacan enjoins her. After all, the technology of the lens, in providing—constructing—the clinical "truth" of the body, enables the body to be gendered by an all-determining relationship to the phallus, its presence or absence. Only under these historical circumstances does it make sense to think in terms of the female as fetish, at least as constructed within an accomplished theoretical position. The "History Portraits," with their plays at gender mutation, seem to assert that to read past acts of gendering one must also read against the grain of the present that they inevitably—inexorably?—also must represent. In so doing, the portraits suggest that Sherman temporarily summons up a time-before: before the castrated woman, before the consolidated discourses of subjectivity and interiority, before the technologies of truth that she burlesques in her corporeal motility. A time, it may be said, before the dematerialized but self-knowing Cartesian subject, or the post-Freudian subject of lack, held full sway.

In fact, the dazzling and grotesque play of surface foregrounded by the prostheses accompanies some evacuated faces—or faces, to put the case more usefully, whose task does not wholly seem to be to create the illusions of a rich interiority for public consumption or viewer interpellation. But this refusal to model a depth model of representation does not converge on another discourse of loss, a loss of the past, to which images devoid of presence constitute a desperate and fetishistic link: they are not, as I have already suggested, Jamesonian. After all, Sherman is in some important sense "there": owing to her prior work of self-representation, one might say her presence is recognizable *because* it is transformed, because, that is, her image foregrounds both the necessity of self-alienation and the inescapability of her present construction as a subject. Given that these photographs are nothing so simple as "self-portraits"—given, in fact, that their avowed sitter is "History"—they may not simply dramatize a ludicrous failure to intersect with the past. Instead, they are invitations (witty ones) to consider a way out of the canonical reconstruction of all species of history, which demands a seamless interpellation, an identity between past and present. As I have suggested, it is just this structure of identification—which could also be termed exemplarity—that is the legacy of humanism institutionalized, which has determined how literary practitioners construct history.[8] Sherman's visual puns on anachronism, her overt forgeries of the subject of Renaissance painting, provoke the reader of Renaissance texts to consider the material freight of the past.

What the "History Portraits" offer in place of humanist identification is something akin to the Foucauldian project of genealogy. The past must be read as radically different from the present moment of reading; but equally, that reading must always and overtly be compounded with all the categorical interests and investments that cannot but be brought to a retrospective hermeneutics.[9] In this light, it may be of interest that, with few exceptions, the styles Sherman imitates are not the most famous masterpieces of Italian Renaissance painters; only the homoerotic Bacchus by Caravaggio insists on identifying itself as an exact citation. In denying herself the right (which is really a constraint) to inhabit the Mona Lisa, for example, Sherman broadens the horizon, extends what must be taken into account when we consider the hallowed and canonical representations of the past, the burnished exemplars of which hang on institutional walls, or gain familiarity through endless reproduction. And what must be taken into account here is corporeal difference, the intransigence of the material body that has since the early modern period been idealized and effectively shunted away from the multiple sites of engagement that determine the modern world.

Sherman provides a critical model for examining the legacy both of a humanist inscription of the past and of an alternative model of past relation that can, for lack of a better term, be called protoethnographic—imaged, respectively, by her investment in art history and by her gleeful rendering of even the Europeanized social body as Other. The models of pastness that her "History Portraits" play on share one important trait: whether exemplary or estranged, each constructs a past apprehensible because of a suppressed act of symbolic violence, one that, if nothing else, has demanded an increasing disembodiment for

the accomplishment of its primary subject-position. How that disembodiment is effected both within the exemplary philology of Renaissance humanism and the materialized objectification of colonialism
and science remains to be seen.

The critical rediscovery of the body as a site of representation and
theory, which Sherman's stills have highlighted, curves back to the
second set of images, the engravings done by Theodor de Bry (from
watercolors by John White) for Thomas Harriot's 1590 *Briefe and True
Report of the New Found Land of Virginia*. One of these images shows a
weroan, or "great Lorde of Virginia," among the Algonkians (figure 4);
the second, an oddly similar representation of a Pict, the authority for
which is "assured in an oolld English cronicle" (figure 5).[10] Like the
other tribal embodiments of English antiquity to be found in de Bry's
images, the Pict is appended to Harriot's *Report* on North America in
order to compare one place to another. The captions aid in this project,
explicitly linking the Algonkian to the Pict, yet simplifying the complex, and contradictory, representational codes operating in the engravings by recasting them as the ideological propositions of early modernity. The barbarity of the early British is very much to the point, as the
text avers, "for to showe how the Inhabitants of the great Bretannie
have bin in times past as savage as those of Virginia" (75). And, if
anything, the Pict could be deemed even more savage than his New
World counterpart, given the trophies of decapitation he brandishes.
Further, the extensive body decoration he sports in lieu of clothing
renders him in some ways more "exotic," more estranged from the
bodily decorum of the late Renaissance, than the comparatively modest
daubings and drapings—and the titular nobility—of the Algonkian.[11]

Despite the apparent clarity of the historical reasoning he embodies, this fantastic British native delivers a mixed message to late
Renaissance readers of the *Report*—as well as to latter-day spectators of
the work of Cindy Sherman. Although the Pict is brought in as a
proximate avatar for the precursors of the English race, his temporal
distance from any site of contemporary reading seems as significant as

A weroan or great Lorde of Virginia. III.

He Princes of Virginia are attyred in ſuche manner as is expreſſed in this figure.
They weare the haire of their heades long and bynde opp the ende of theſame in
a knot vnder thier eares. Yet they cutt the topp of their heades from the forehead
to the nape of the necke in manner of a cokſcombe, ſtirkinge a faier lōge pecher of
ſome berd att the Begininge of the creſte vppun their foreheads, and another ſhort
one on bothe ſeides about their eares. They hange at their eares ether thicke pearles,
or ſomwhat els, as the clawe of ſome great birde, as cometh in to their fanſye. Moreouer They
ether pownes, or paynt their forehead, cheeks, chynne, bodye, armes, and leggs, yet in another ſorte
then the inhabitantz of Florida. They weare a chaine about their necks of pearles or beades of cop-
per, wich they muche eſteeme, and ther of wear they alſo braſelets ohn their armes. Vnder their
breſts about their bellyes appeir certayne ſpotts, wheat they vſe to lett them ſelues bloode, when they
are ſicke. They hange before thē the ſkinne of ſome beaſte verye feinelye dreſſet in ſuche ſorte, that
the tayle hangēth downe behynde. They carye a quiuer made of ſmall ruſhes holding their bowe
readie bent in on hand, and an arrowe in the other, radie to defend themſelues. In this manner they
goe to warr, or tho their ſolemne feaſts and banquetts. They take muche pleaſure in huntinge of
deer wher of theris great ſtore in the contrye, for yt is fruit full, pleaſant, and full of Goodly woods. Yt
hathe alſo ſtore of riuers full of diuers ſorts of fiſhe. When they go to battel they paynt their bo-
dyes in the moſt terible manner that thei can deuiſe.

4. Theodor de Bry, "Weroan, or great Lorde of Virginia," from Thomas
Harriot, *A Briefe and True Report of the New Found Land of Virginia* (by
permission of the Folger Shakespeare Library).

25

5. Theodor de Bry, "The true picture of one Picte," from Thomas Harriot, *A Briefe and True Report of the New Found Land of Virginia* (by permission of the Folger Shakespeare Library).

the common ground upon which he is purported to stand. The opposite is true for the Algonkian: while he shares the temporal frame of those who might gaze upon his reproduced image, he is distanced as much by the status of quasi-primitivism attributed to his culture as by geography. Seen from the vantage of the seventeenth century, the English "then"—when savage Picts were the prime inhabitants of Britain—become the (North) American "now," and the comparison one of time as much as of space. One might contrast this representation of the British self as formerly "other," barbaric, with an alternative, even classical, model of Englishness. As Spenser and others demonstrated, this history could bypass the anthropological model offered by the New World, to be imaged instead through a literary lineage extending straight from the Trojan Brutus to the Renaissance English.[12] These competing exemplars of past humanity, the quasi-classical and the quasi-ethnographic, begin to hint at a point to be taken up more extensively later: that the past is made, not found—produced according to the complex interplay of archival information, the rules of nascent discourses, and the exigencies of ideology.

The temporal opposition set up by the illustrations to Harriot's *Report* is one instance of such an interplay, and its effect is to construct useful resemblances over both time and geography. When the salient, specific aspects of geography, as I will argue in connection with Galileo, come to be seen as "space," there has occurred an evacuation of the local and contingent. Another useful isomorphism thus emerges—between the manner in which the alien landscape of Virginia becomes both domesticated and universalized, and the regularizing of the material world to be found in the emergent practices of seventeenth-century science as represented by Galileo. In the correspondent play of images found in the Harriot text, the agenda that underwrites the juxtaposition seems familiarly, because universalizingly, ethnographic: two different objects of knowledge are put together, the better to establish a relational link between them. Printing the Pict after the Algonkian suggests that radically differing bodies have become equivalents: indeed, the common modeling of the bodies—their conventional, highly

stylized musculature—is the sign of that equivalence. By logical extension, the cultures that the Pict and Algonkian represent, serve as points
of entry for, are brought into a contiguous narrative space.[13] But not, of
course, to argue for anything like their equivalent cultural value at the
moment of a contemporary seventeenth-century reading, since the juxtaposed illustrations use an ideologically loaded notion of temporality
to deny the non-European specificity of the Virginia Indian and to
reposition him in relation to European history. Their common musculature is, after all, coded through the European aesthetic tradition. If
the differences between the one and the other, Pict and Algonkian, are
made available and reduced to the visible by these engravings, then the
difference between them is a matter of time. "Time" must therefore
function to signify in a newly interested fashion. It serves at once as
a measure of linear regularity, and a guarantor of relation even where
no prior train of geography, nationality, or other form of consequence
exists.[14]

    "Old World" and "New," relation and alterity, are mutually con-
stituted as modern formations through a fantasy of development that
itself positions temporality as a neutral and universal substrate. Yet
the plotting of bodies and cultures along a one-dimensional time line
marks the temporal as a loaded and interested element in emergent
forms of power-knowledge. As the anthropologist Johannes Fabian has
noted of the early history of his own discipline: "what could be clearer
evidence of temporal distancing than placing the Now of the primitive
in the Then of the Western adult?"[15] Although Fabian is speaking of
ethnography as an accomplished, post-Enlightenment epistemology,
the problem he makes of "temporal distancing" also illuminates the
hidden presumptions that inform colonialist discourse almost from its
inception. The images of the Algonkians consolidate "Englishness"—
and, by extension, Europeanness—at the same time they appear to
embrace a universalizing narrative of development. That the consolidation of identity is achieved by a recourse to alterity marks the *Report* as a
protomodern document. The twin assertions of distance and related-
ness offered by words and pictures in Harriot propose a preliminary

version of modern disciplinary power—the power to make objects. Hence, this early modern modeling of human (ethno)history is not the pliant and playful "rehearsal" of culture for which the Renaissance has been a recent referent, but the making of one culture through the positioning of another as a site of scrutiny.[16]

During the period being considered, the ideological distancing provided by narratives of parallel but unequal development in New World and Old guarantee that the present moment of European history is valorized over a past that begins to be labeled as immature, inferior. The deployment of temporal relation, if taken on its own terms, has much in common with the reinscription of classical time offered by earlier Renaissance humanists, as the next segment of this chapter suggests. But unlike that other backward-looking formation, the protoethnographic modeling of the past suggested by Harriot's *Report* disables the past of the authority granted to it by virtue of its very pastness, its maturity, its venerable—because classical—antiquity.

That the past could be authoritative, indeed, be considered the parent of the present, is a salient feature of Renaissance humanist culture: reverse the sign of the past, and what emerges is time as a symptom of nascent modernity. To evaluate the competing ideological work done by Harriot's modeling of a so-called primitive past, then, it is necessary to understand the text-based investments in classical antiquity associated with Renaissance humanism. That phrase, like most other efforts to fix a historical formation for use in argument, covers over much that is contradictory and much that is in need of further elaboration. Nevertheless, there are certain aspects of humanist culture that can be selected out from its repertory to provide an epitome—a schematic distillation that seeks to uncover what may be at stake when early modern European culture constitutes itself against the authority of classical antiquity.

My argument here both reaffirms and recasts one canonical description of Renaissance humanism: that it saw the emergence of what can be called historical consciousness—a consciousness that the past was importantly different from, and discontinuous with, the present; and that certain discursive consequences follow from that emergence.[17] By stressing the importance of historicism to the formation of humanism, it becomes possible to acknowledge the ideological valence of humanist philology, its immanent assertion that classical texts are repositories of social value. Humanist reading and writing are crucial evaluative activities, crucial as well to constituting the early modern (male, European, scholarly) subject. When a humanist scholar reproduces, imitates, annotates, or merely studies Roman rhetoricians or Greek poets, he in effect identifies with them to assume their virtues as a function of their language, and so he models his subjectivity in relation to theirs.

Although such a structure of identification is not itself isomorphic with the latter-day bourgeois practices against which I have read Sherman's photographs, even at this formative stage Renaissance humanist identification is not without its exclusions. Indeed, the humanistic identification with past subjects makes homologous signification a privileged bridge to bygone mentalities: as valorized by its philology, the knowledge of classical Greek, and, especially, Latin, collapsed the temporal distance separating the recuperated texts of Greece and Rome from the vernacular culture of Europe. To put it more simply, humanist learning augmented the distance of time with the contiguity of language. The closer a text could get to Cicero's Latin the better, even to the point of parodic contortion—as Erasmus's *Ciceronianus,* a satire against the many slavish imitators of Cicero who would only use the verb *tenses* to be found in the master, demonstrated to witty effect.[18]

A less drastically mimetic example is provided by Petrarch, who is generally considered the first Renaissance humanist and whose relationship to the foundational texts of classical antiquity might thus be claimed as formative. When he collected his familiar letters together in imitation of Cicero, he included epistles to such classical authors as Horace, Livy, Quintilian, and the inevitable Tully. Such a collocation made the Italian poet's letters a site for the invention of "familiar" consciousness—the conversation, if not of coevals, than of folk who by virtue of a shared language have a lot in common. In one letter to a contemporary, in fact, Petrarch insists that "Cicero" himself has wounded him grievously—Cicero, that is, as personified by a volume of his texts, which had fallen from a shelf and bruised the poet's leg.[19] The conceit participates in consolidating an important truth: in Petrarch's phantasmatic collection, the correspondence that for Foucault in *The Order of Things* was the epistemic foundation of the Renaissance appears to govern the erasure of diachronicity. Resemblance is both the order of, and the ordering of, the day.[20]

Even the quickened interest in exposing—and manufacturing—forgeries characteristic of Renaissance philology arose out of a complex attempt to re-create, reinhabit, the past through linguistic and historical knowledge. In this regard, consider Lorenzo Valla's influential detection that the "Donation of Constantine"—a document that ceded imperial lands to the papacy—was a fake, based on the humanist's authoritative examination of the Latin in which the document was composed. Valla's treatise betrays an exquisite and scholarly sense of the mendacity, the will to deceive, that is unwittingly revealed by the telltale linguistic anachronism into which the forger blunders. Possessed of the historical understanding of language that the Donation's forger so crucially lacks, Valla appears to be everything that Petrarch is not—distanced, measured, far from captive to immediacy. Yet Valla's predication of late antiquity as a time of perceptible difference, a difference moreover that can be recaptured through critical analysis, makes his treatise not so very different from Petrarch's familiar letters.[21]

Seen with a slightly different turn of the lens, Petrarch's epistles offer to make up for the same loss that a forger attempts to remedy. After all, most modern inscriptions of the past, whether in material or discursive practice, are involved with concepts that forgery particularly foregrounds: knowledge, authenticity, identification, the accuracy of a historical empathy. Anthony Grafton has accordingly written that Renaissance forgery arose, not simply to question or defend territorial or property rights, but also to give rein to a more complex nostalgia for the productions of the past:

[Forgery] aimed above all at recreating a past even more to the taste of modern readers and scholars than was the real antiquity recovered by historical scholarship. Many of the early recorders of monuments and inscriptions filled in missing texts in their notebooks just as they would the missing limbs and heads of statues, moved by an exuberant desire to see the ruined past made whole again.[22]

When the fake artifact from the fake past is exposed, certainly it may also expose the impoverished or incomplete technique of any individual forger, owing to such factors as the default of technology. Hence Valla's ease of scholarly detection. But the failed forgery is more generally revelatory, since it more generally discloses the defining horizon within which historicist practice takes place. Rather than think of forgeries as an aberrancy, a corruption of an otherwise stable system of truth and value, perhaps they are an inevitable consequence of the desire to enter into and partake of the past that systematically emerges in the early modern period. What I have termed humanist history, for instance, effaces the signs of its contingent production, the better to provide an exemplary and convincing rendition of a remote time, of a past into which a reader can insert himself because the historian—like the successful forger—has already done so.[23] The moment when the historical narrative fails to sustain (and so needs augmentation or supplanting) is not unlike the moment when the forgery is exposed: a crucial lack is opened up, an absence of identification and sympathetic imagination—or, as Sherman's fake Old Masters suggest, a failure of conviction shared as readily and naturally by those who consume the past as a source of value as by those who instantiate it.

Thus Renaissance forger and humanist critic were often the same person. The careful scholar who argued for the inauthenticity of one document might well author another much like it—taken, this time, by contemporary readers to be a work of classical antiquity. Grafton has spoken of a (non-Freudian?) nostalgia for "wholeness" betrayed in the repair of statuary and the filling in of historical blanks. Yet this apparently indiscriminate desire to reconstitute the whole is perhaps not as self-evident as his words make it appear, not merely a desire to reverse the iconoclasms of time. The present of Renaissance humanism is apparently permeable to the past, bespeaking not so much Grafton's "exuberance" as an incomplete—emergent—sense of historical separation as a definitive constituent of modernity. In fact, seen from this position, humanism becomes a willed oblivion to the conditions of historical difference, which becomes rearticulated in the abstract universals of liberal humanism.[24]

In fact, the connection between the humanism of the Renaissance and its later, post-Enlightenment avatars is not far to seek. While these later versions are materially imbricated in the overt imperialism of the European nation-state, the voluntary (inter)nationality of linguistic isomorphism among Renaissance humanists figured a different sort of political federation, a *res publica litterarum*. Although a state without material borders, in effect a state of mind, this republic of letters comprised an elite citizenry founded on a certain type of literacy—a skill in reading, inhabiting, and reproducing the texts of classical antiquity—that prefigures the systematic education provided to indigenous peoples by Catholic missionaries or state-sponsored colonial bureaucracies.[25] Most importantly, the imitative practices so important to Renaissance humanism—the effacing of temporal and geographical remoteness in the name of a language presumed to be universally apprehensible, like classical Latin—underwrite an ideologically interested similitude among "men," whose infinite glory and upwardly mobile intellects were therefore to be celebrated.

Of course, the sign of "man" is universal only within the conservative dispensation of humanist culture. As Juliana Schiesari has shrewdly observed, "Humanism's praise of the 'dignity of man' appears predicated upon the abjection of what it considers the 'non-human.' "[26] Men, both ancient and modern, whose sense of human history and possibility was forged through Latin and Greek were the privileged subjects of humanism; all others—artisans, women, the inhabitants of the New World—need not apply. Much follows from that exclusion.

One thing that distinguishes Renaissance humanism from later formations, however, is the comparative humility of its practitioners in the face of classical knowledge. Although their culture is enriched because it has annexed the intellectual property (what may be termed anachronistically the cultural capital) of Greece and Rome, Renaissance humanists initially conceived of themselves as the junior branches of this incorporated culture. In his epistle to Homer—written in response to an actual letter from a friend (an act of ventriloquism that shows how permeable the boundary between classical antiquity and the fourteenth century could be made to seem)—Petrarch positions himself literally as the *infans,* deprived of language, eager for nourishment: "Yet it is with me as with a babe: I love to babble with those who feed me, even though they are skilled masters of speech."[27] According to the logic of generation and development that informs Petrarch's response, classical antiquity represents the maturity of European—which is to say, world—civilization. The moment from which the Italian humanist speaks is the product of that maturity, but a juvenile one. The difference in sophistication and development between his present and the past of Cicero and Homer is represented by a familial cultivation of speech that serves as a domestication of humanist philological practice.

Yet the humility to which I have referred does not represent the whole picture. Nor could it, if we remember the project of de Bry's illustrations, which valorize present-day Europe over the primitive past of the native inhabitants. What remains to emerge in early modernity is the value of contemporaneity, the worth of the present moment. Contrast Petrarch's touching faith in the benign authorities of his classical *patres* with the rather more dismissive characterization of antiquity articulated by Francis Bacon: in his Preface to *The Great Instauration* (1620), Bacon speaks of "the wisdom . . . derived principally from the Greeks [that] is but like the boyhood of knowledge, and has the characteristic property of boys: it can talk, but it cannot generate, for it is fruitful of controversies but barren of works."[28] The Baconian quotation inverts the familialism of Petrarch and puts in question the status of philology as the privileged basis of knowledge. Classical antiquity is not the parent of the present moment, but instead its infancy, and

34

authoritative eloquence as the generator of culture is reduced to unproductive childish prattle.

Bacon, like de Bry, offers an important hint for thinking about how the past becomes deprivileged, becomes but one object of study among many in the modern faculty of disciplines. In imagining that past as embryonic, insufficient, these texts suggest that the classical interpellation of the humanist subject must compete with, and be complemented by, a distinctly presentist subjectivity. But a curious ellipsis, most visible in Petrarch's simile, binds the three (otherwise discrete) instances I have introduced. What happens to gender when the speakers and feeders, the sources of language and knowledge, are the writers of classical authority, as they are for Petrarch? Or are reduced to impotent boys by Bacon's version of humanism? These are, in effect, versions of a question already encountered: How is modernity produced as a consequence of a polarity between the supreme male European subject, whether forward- or backward-looking, and the Other against which it launches itself?

Petrarch's comparison of himself to a baby in need of speech and food—both are coextensive—would seem, in its cozy familialism, to demand a maternal presence. Such a presence, however, is unlikely in the text-based formation that he inhabits, for reason more than the historical fact that comparatively few women had access to humanist education. Although it is too simple to say that Petrarch's ideal family contains no place for the female because it is an ideal family, a disembodied one, this formulation provides a good first approximation for the status of the feminine. A more exact understanding of this "disembodied family" is afforded by the exemplary subject-formation made possible by the humanist program of reading. In choosing among classical texts to read and study, the humanist in effect chooses his relations, scripts himself in relation to canonical forebears in a transaction normatively and unexceptionably gendered. To the extent that gender difference emerges within Renaissance humanism, it is as a tacit substrate, a quasi-material ground of discourse, as chapter 2 suggests more fully. Hence, the unexpected contiguity between the humanist's narrative of linguistic development and the less domestic one enacted within Harriot's text: in both cases the idealizing formation of the early modern subject must do its work in relation to those who for reasons of difference are barred from the project.

Barred—but not wholly banished. As has been seen in the case of the Algonkian, making difference—and making something *of* difference—is the work of modernity in emergence. The same may be said regarding the phantasms of woman in texts like Petrarch's. In fact, the pertinent question may be, not whether women "count," but how to account for the relation between the figure of woman (as over against the body of the Algonkian) and the humanist corpus. Humanism, as a deliberate political culture, figures its legitimacy through imagining a continuity with classical Greece or Rome that suppresses the fact of temporal separation (consider the erasure of the "Middle Ages"). This suppression can be aligned with a more general strategy of domestication available for the institution and maintenance of humanist subjectivity. Thus one critically occluded term in Renaissance humanism, and in the discourses of emergent modernity as well, is the female subject, as Petrarch's *infans* reveals.

The argument can be extended further, into the very constitution of the humanist educational program. As Stephanie Jed has argued in her study of humanist interest in the legend of Lucretia, a reading program based on the classics can constitute a form of symbolic violence to the subject gendered female.[29] This violence is both denied and reproduced, as Jed suggests, by the reiterated dissemination of texts in the Renaissance—when, time and again, the story of Brutus and the founding of Rome is related as foundational for Renaissance polity, for which the rape of Lucretia becomes merely a pretext, a vehicle. Indeed, a gruesome story recounted by Anthony Grafton offers further suggestive evidence of humanism's gender trouble. At the same time that the texts of Cicero were being rediscovered and reproduced, the perfectly preserved remains of his daughter Tulliola were, apparently, often themselves "discovered," again and again for more than a century, at gravesites all over Italy.[30] Grafton offers no further elaboration. Nevertheless, it is tempting to speculate on the efficacy of these female bodies as found relics for humanism's immaterial republic of letters, as material counterparts for the work done by the endlessly reproduced, endlessly defiled Lucretia.

Obviously, no sort of innocence can be claimed as characteristic of humanist immediacy, in contrast to a protocolonialist distance whose dominating effect has been widely recognized. Jed's work usefully repositions humanism's self-articulation, and shows how its collapsing of the distinction between past and present does ideological work in reproducing familiar structures of domination: witness the unhappy proliferation of Tulliolas. Then it becomes useful to compare the multiplication of female bodies in humanism proper with the rather different discourse of the feminine accomplished within Baconian science. In *The Advancement of Learning,* for instance, the relationship between man and nature is explicitly an erotic congress, separated from what is "empty and void," the better to produce "solid and fruitful" speculations: "that knowledge may not be as a courtesan, for pleasure and vanity only, or as a bond-woman, to acquire and gain to her master's use, but as a spouse, for generation, fruit, and comfort."[31] The Baconian program works through the repertory of socially normative relations before seizing on marriage to legitimate its knowledge-seeking agenda in an image that seems at first to suggest the necessary (if, of course, inferior) company of women. Yet the texts's summoning of the scientific subject to marriage with "knowledge" offers merely the tropological appearance of inclusion. Women—and others—may figure "differently," may appear to constitute an element in the universal project of science; here, no less than in humanism, they become the matter to be worked upon, assimilated, domesticated.

As the Petrarchan quotation indicates, the Renaissance discourse of humanism speaks soothingly of family ties, of a fantasy of intellectual nurture in which traditions of gender play no part. The contrast provided by Caliban is well-known: that thing of darkness is taught language in order to know and curse the great Italian father who can barely acknowledge his complicity in the formation of the slave's consciousness.[32] The question of language is thus overtly and crudely political—most critically, perhaps, insofar as it constructs past times in relation to the present, sees prior temporality in terms of a linear, organic development. How is the European past reprocessed, how does the emergent modernity of late Renaissance culture renegotiate its parental thrall to classical antiquity? How did the past come to be undomesticated? The answer, as I have already suggested, requires an Other—an Other, whether Algonkian or feminine, excluded from the full range of humanist subject-positions because excluded from the signifying process. In that all-defining regard, de Bry's illustrated man is not much different from the female corpses Renaissance archaeologists called Tulliola. By virtue of their muteness, each is delimited by the brute fact of embodiment, or else incorporated through reduction to mere tropological convenience.

As an experiment in ethnography before the fact, the *Report* competes with, indeed, comes to supplement as the discourse of the scientific, the reproduction of the Western past associated with humanism to which I have alluded. There, notions of antecedence are imbued with and conveyed by textual presence. Classical antiquity speaks to the Renaissance humanist through a language of contiguity ideologized as transparent to its origins, while the native, whose linguistic competence defies humanist philology and therefore defines its limits, is spoken by his body, his customs, rather than his textual traces. In fact, the "textlessness" of native peoples renders them the readier for absorption into a universalizing narrative of human development. As Samuel Purchas has written, men wanting the "Use of letters and Writing" are "esteemed Brutish, Savage, Barbarous."[33] And, consequently, ephemeral:

by speech we utter our minds once, as the present, to the present . . . but by writing Man seemes immortall, conferreth and consulteth with the Patriarks, Prophets, Apostles, Fathers, Philosophers, Historians, and learns the wisdome of the Sages which have been in all times before him; yea by translations or learning the Languages, in all places and regions of the World. . . .[34]

Purchas's version of humanism extends the temporal mastery of the Renaissance subject by means of the material canonical text. But the lack of such texts places North Americans like the Algonkian in a temporal suspension, a time without a prior inscription, evacuated to suit the ideological needs of European protomodernism.[35] Like Cicero's daughter, he must be understood by corporeal analogy, by the representation and examination of his body.

The illustrations for Harriot's *Report* convert alien beings into evidence, render cultural difference available for assessment, at the same moment that the contemporary European male body is exempted from scrutiny and removed, in and by theory, to the site from which the possibility of assessment is contemplated. De Bry's versions of the drawings by White provide no depiction that commemorates European interactions with the inhabitants they observe, no place that the Algonkian can be seen to share with the contemporary English explorer. It is as if the explorers, in contemplating their relation to the Virginians, occupy no space at all—which is, in a sense, the province of theological afterlife as much as it will come to be of science.

The close kinship between the dematerialized subject, as it emerges in science and in the colonial enterprise, and its prior theological antecedent is best demonstrated by John Donne's 1611 polemical treatise *Ignatius His Conclave*.[36] Donne's tract lends itself to a symptomatic reading like the ones that have arisen from Sherman and de Bry. Its articulation of a demonized novelty within the conservative parameters of Protestant theology (and the image repertory of Catholicism) produces novelty as a historically inflected cultural artifact. "The new"—change recognized as without precedent, without (classical or divine) authority—becomes a marker, a semiotic flag making it possible to chart the emergence of the early modern world from within Renaissance culture, and scientific modernism from humanist antiquarianism.

It has generally been suggested by Foucault that many of the institutions and formations, the conveyors of objects, disciplines, and forms of attention, characteristic of Western modernity have genealogical antecedents in the religious exercises of post-Tridentine Catholicism. Donne's text, as (among other things) an anti-Jesuitical polemic, does not readily offer itself as evidence for preemergence. Yet it seems possible to suggest that, while far from Catholic in its thematic material, *Ignatius* bears some resemblance to the exercises in meditation whose presence Louis Martz once identified in seventeenth-century English literature.[37] Indeed, the starting point in the text is an "extasie," a state of sustained and altered spiritual transcendence not unlike that generated by Catholic meditational exercises. From there it moves into something like a travel account, complete with an ethnography of hell. The movement to spiritualized disembodiment allows the observer not only to explore hell, but to put its principal occupants, Lucifer and St. Ignatius Loyola, under surveillance, under a gaze equally mocking and inquisitive.

As I have already suggested, the beginning of Donne's text problematizes the connection between a spiritually inflected disencumbrance from the body and the dematerialized viewer of early modern science and exploration. Michel de Certeau has examined mysticism—spiritual experience likewise figured through flight and bodily disencumbrance—in ways that further heighten the connection to emergent modes of modernity. Especially in the sixteenth and seventeenth centuries, its "proximity to a loss . . . [a]t the dawn of modernity" causes mystic experience to search for an explanatory system linking the universality of absolute divinity with the particularity of the individual subject.[38] Such a quest can be seen as isomorphic with the general explanatory schemes, the interplay of abstract and particular, that constitute emergent science in the period. Consider in this regard that the privileged observer of Donne's hell is presented as questing, curious, and disembodied himself; his ecstasy grants him the privilege to roam the heavens, a privilege he explicitly compares with those assumed by Galileo, Kepler, and Tycho Brahe (7). Later, he speaks of being "hungerly caried, to find new places, never discovered before" (9). It will perhaps be no surprise, then, that among those he subsequently identifies as seeking admission to Lucifer's domain are Columbus and Copernicus. From the outset, the critical observer seems, both in his range of associations and his affectation of position, to already have been conjoined with those whom he observes doing the devil's work, the bringing-to-being of modernity through innovation. Here, as in the case of Harriot, materiality and its effects are displaced onto the bodies and spaces to be observed and subjected. Hell, in fact, becomes a limit-case for the fantasy of disembodiment, infinite movement, and om-nivoyance (if seeing becomes the sign of knowledge, then to see all is to know all), that, when legitimized through scientific practice, will sig-nify the ideology of the new.

Consequently, this initializing hell is a site of peculiar constitu-ency, its inner precincts the desired residence of the most innovative. As the visionary, voyaging narrator describes the scene, "onely they had title, which had so attempted any innovation in this life, that they gave an affront to all antiquitie, and induced doubts, and anxieties, and scruples, and after, a libertie of beleeving what they would; at length established opinions, directly contrary to all established before" (9). Donne presumably wrote the text as part of the pamphlet wars between James I of England and the Roman Cardinal Bellarmine concerning the Catholic assertion of papal hegemony over political authority, and that occasion accounts for, although it does not exhaust, the prominence of Ignatius Loyola as enforcer of hell's door policy. The Jesuits, established by papal bull in 1539, were themselves a fairly recent religious order; in addition to the specific polemical target Ignatius provided Donne, the newness of the order he founded marks him, too, as an agent of "the new."

In fact, Donne's text suggests a generalized anxiety about novelty, about promiscuously burgeoning forms of cultural and material pro-duction in the early seventeenth century. Those who come to throng for admission include inventors of "new attire for woemen . . . of Porcellan dishes, of Spectacles, . . . of stirrups" and of importers of "Caviari" (65); as I have already noted, they also include Copernicus and Columbus, as well as Machiavelli, Aretino, and Paracelsus. Canonical agents of his-torical change are jumbled with the anonymous agents of mercantile capitalism, and the jumble itself attributable to the logic of the market. In effect, everyone is trying to buy his way into Satan's inner sanctum, whose exclusivity Loyola zealously maintains.[39] This more "inward" place in hell betokens modernity as a zone of exemption. Equally free, it seems, from prior discursive constructions and from theological traditions concerning the damned, Lucifer's space thus becomes the most privileged domain of those who would be accounted cultural innovators.

Donne's polemical text does the obvious work of conservatism, as Marjorie Hope Nicholson suggested.[40] Specifically, it attempts to contain the semiotic agents of early modernity within a recursive, eschatological notion of time, to deny innovation, and hence a history recognized as such, by inserting them into the overarching scheme of retributive providence. Change becomes sin, linear time a moral dead end. But Donne's hell is less an afterlife, a space, that is, after history, than a symptom of the inescapability of this new version of history, one that renders the obverse of the familiar Renaissance equation between the "New World" and a visible paradise. Here, it is new world as hell. Donne's text partakes in what it demonizes; it is itself a production of the new and cannot escape its own historicity.

To interrogate Donne's metaphysics of modernity, which produces the modern, authorizes it, even as it traduces innovation, is to rest heavily on its tacit investment in materiality, as signified by the amassment of new objects and devices that are so demonized. But it is also to invert its demonizing of the new and to ascribe a counterforce to the work of culture *Ignatius His Conclave* tries to suppress. Elsewhere, in other cultural spaces, it is not the new that is decried but the old. A concrete instance of how newness—exploration, investigation, transgression—is valorized can be found in Carlo Ginzburg's study of the changing fortunes of Icarus in sixteenth- and seventeenth-century emblem books.[41] From a graphic representation for foolhardy, even impious ambition, Icarus's flight becomes the icon of a transgressive audacity, an audacity that equally ratifies the material quest to go beyond—in voyages of trade and exploration—and the intellectual one of which it is an overt sign. The glorified career of Daedalus's son signifies

42

the jettisoning of theological encumbrances whose belatedness is inscribed across Donne's text, to provide moral ballast against the entrepreneurial and discursive ambitions of modernity. Indeed, flights of intellect and fancy prove hard to police, even in texts that, like *Paradise Lost,* aim to legislate against such mobility and such access to power. As chapter 4 shows, Raphael's injunction to Adam to "be lowly wise" (8.174) is contradicted by the very attention the angel gives to Adam's intellectual curiosity about the heavens; more drastically, perhaps, it is also contradicted by Satan's compelling flight through the "Space" of Chaos, in search of the "new Worlds" (1.650) that it is rumored to produce.

The emergent formation which glorifies a sovereign, even liberal model of subjectivity offers a challenge to the authority given by humanist book culture to the classical past and the antique. The model of humanism, of contemporary civilization founded on an unmediated dialogue with classical antiquity, is supplemented by an emergent sense of the past as encumbrance, a supplementation that reverses the vectors of authority which originally positioned such a resuscitation. As the Bacon citation from *The Great Instauration* suggests, "production," whether in the familial mode or the mercantile one, is the province of new systems.

However, new discursive relations can also be worked out through old things, in sites already inscribed by dominant formations. Just as Sherman's photographs offer her body as the site for her engagement with historicity, the recalibration of cultural value I have referred to is accomplished in the world of the material, the object, as well as in the dematerialization of the subject. So much is clear from Donne's equal distaste for philosophical innovators and novel consumer goods. The material negotiation in the relative worth of the new versus the old finds a particular resonance in the early modern collections of memorabilia and artifacts variously called curiosities, *Kunstkammern,* or wonder-cabinets, whose discursive importance in the "rehearsal of culture" has become well-known. Steven Mullaney argues for the essential strangeness, alienation, of the early modern collection in comparison to the model of the modern museum: "Its relation to later forms of collection is a discontinuous one, even when the objects displayed were themselves preserved and carried over. . . . The museum as an institution rises from the ruins of such collections . . . it organizes the wonder-cabinet by breaking it down—that is to say, by analyzing it, regrouping the random and the strange into recognizable categories that are systematic, discrete, and exemplary. The museum represents an order and a categorical will to knowledge whose absence—or suspension—is precisely what is on display" in the wonder-cabinet.[42] Seen in this light, however, "rehearsal" offers a model of cultural practice that is already divorced from social significance—the final chance for objects to play before they get harnessed to the real work of modernity.

Rather than view the wonder-cabinet as quaint, aesthetically irrelevant, or simply estranged, it might be better to consider it as a protoinstitutional site of epistemological contestation. Hence the collection stands as an agent of modernity as much as its apparent randomness might be a casualty of it. Eilean Hooper-Greenhill, for instance, has argued that the Renaissance epistemes of correspondence and resemblance constitute the historical a priori for the proximity of objects in a collection; the cabinet of curiosities was a *theatrum mundi,* a viewing place for the world.[43] When meaningfully sampled through its artifacts, the world, understood as total cosmological system, can then be known through a magisterial survey, through a visual itinerary that takes past and present, far and near, into its purview. The cabinet artifact thus plays a substantially different role from the one it comes to occupy in succeeding epistemes. When modern museums set out objects of knowledge, the principles of classical taxonomy and hierarchizing orders of knowledge embedded in its practice demand the subsuming of materiality within representative discourse. What may be seen in later museum practice is the power turned over to the conventions of professionalized installations by viewers who seek out wall labels and other explanatory paraphernalia.[44] In contrast, Renaissance cabinets constitute a prior discourse of materialization isomorphic with that effected by the increase in printed material after the invention of movable type—which placed in texts what had before been confined to memory. If, as Hooper-Greenhill suggests, cabinets are a triangulation of the emergent print culture of the late Renaissance with the classical "theater of memory," then the collection becomes in effect a visual encyclopedia, "an example that encompassed both the space of the library and the space of the theatre" (102).

It has been suggested that wonder-cabinets emerge in the late Renaissance from personal collections of classical antiquities—such things as capitals, coins, and friezes—and expand, with time, to include bits and pieces of the exotic, the New World and the natural world—deformed births (as in Ambroise Paré) and other remarkable human specimens, but also plants, animals, and minerals.[45] That hypothesized point of origin is as important as the subsequent career of the collection as a model for the study of human history.[46] In the prototypical wonder-cabinet, the classical past of humanism is seen as coextensive with its fragmentary material trace and compacted into it, and installed in a space of equivalence with the evidentiary oddments of colonial ventures and protoscientific ambitions. When actual artifacts could not be obtained, representations constituted acceptable substitutes: thus, the painting *Amerika* by Jan van Kessel, which depicted Brazilian tribal figures settled in a Baroque chamber full of curiosities.[47] When denatured and set within this purview, "the past," so constructed, is clearly less a source of authority than a demystified relic, an object of studious curiosity rather than a site of primary investment. The "antiquities" displayed in these collections are not antiquity as understood by the early Renaissance humanist, and what is to be learned from them is not identity; perhaps, however, it is "history" in the emergently modern sense of the term.

These wonder-cabinets partake in the dynamic both of the souvenir keepsake and the collection, the part and the (supposed) whole. Susan Stewart has spoken of souvenirs as metonyms that, like the forgeries Grafton has written of, betoken a site of nostalgia. From a phenomenological standpoint, these objects are concretions that both remind one of something that exists or has existed in another time or place, and the impossibility of ever retrieving that thing, past, or place in its authenticity. Thus, the metonymic object is a sample, in material form, of the

now-distanced experience, an experience which the object can only revoke and resonate to, and can never entirely recoup. In fact, if it *could* recoup the experience, it would erase its own partiality, that partiality which is the very source of its power. . . . the souvenir must remain impoverished and partial so that it can be supplemented by a narrative discourse, a narrative discourse which articulates the play of desire.[48]

The desire mobilized in collecting, when understood in historical terms, is equivalent to that revealed by Valla's anxiety over the linguistic anachronism, the fake from the fake past: to have the antique past whole and present, to inhabit the pure space of correspondence that Petrarch's letters seek to bring into play. The narrative of the souvenir is effectively that of the humanist project.

Collections of the sort I have been describing, however, insist increasingly on the past as a complex production: of course, it still remains the familiarly classical past of Petrarch, but increasingly it is also reconfigured as the exotic, the primitive, the curious. The ever-greater number of "other" objects in the wonder-cabinet allow for the possibility that the primary humanist model of the classical past has been deaccessioned, changed over from source of authorizing value to an object of study among many. The juxtaposition of classical fragments with the instruments of modernity, as signified by globes, lenses, microscopes, and botanical specimens, reproduces a moment of flux, where once-dominant forms of apprehending European culture compete with emerging forms.[49] Concomitantly, the "transgressive" New World, as represented through its objects, could be brought into an equivalent knowability, an equivalent domestication, in the magisterial space of the *theatrum mundi*.[50]

These passes at ethnography, via the wonder-cabinets and the engravings, imagine how the antique and the primitive might be reprocessed under the sign of an emergent modernity, to be hypothesized as in need of domestication in discourse. So differential a reading of the past, which might be understood as an allegory of origins for the human sciences, is enabled by its encounter with alterity. As I have suggested, my rhetorical reversion to materiality is a way of rendering for inspection a set of cultural changes that, given their interstitial nature, might not otherwise be capable of apprehension before they come into full visibility with the Enlightenment.

This recourse to wonder-cabinets, to metonymic samplings of the world, has itself been introduced as metonymic for the transcoding of materiality that takes place with the emergence of modernity. It is also a way to make palpable the relation between humanism and modern structures of power-knowledge when those modern structures begin to emerge in late Renaissance culture. This relation has been defined in terms of a simple opposition, as the wars that from the vantage of a celebratory modernity have been codified as the debate concerning the ancients versus the moderns.[51] However, given the complex discourses around temporality itself that I have been examining, I do not want to reinscribe the work that has already been done but to reconfigure it, to add a third perspective to the relentless binarism of then-and-now. That third perspective is a way to unconceal the shifts in culture and power that underlie the growing investment in modernity, even in this early and incomplete form, by asking how emergent scientific ideology, as a structure of feeling immanent in texts, was assigned productive force in the sixteenth and seventeenth centuries. How is temporal relation made responsive to the exigencies of the emergent ideology of science? What is the place of the "New World," and of the mechanisms of difference made visible through colonialist discourse, in this transfer of cultural authority from old to new? One last question, at first put simplistically: what is the place of "literature," constituted as an object of study, in among the emergent faculty of disciplines? The question should better be framed with more historical nuance to consider how the processing of European discourse through a site of difference maps onto the inhabitation of scientific representation from narrative forms associated with humanist literature, such as the *Utopia*.

The following chapters, which use exemplary readings of "literary" and then "scientific" texts to chart a shift toward the forms of knowledge valorized by modernity, offer the best answer to the questions I have asked. But the questions themselves are predicated on an understanding that the forms of narrative representation now named as "literature" were coming into discourse during this period and that an appropriate regime of truth was being delimited around the discourse by humanist critics. Against this regime of the "truthful fiction" a counterregime was being deployed, most evident in the epistemological contestations that focused on astronomical hypotheses. The "literary" thus came into problematic visibility during the sixteenth and especially seventeenth centuries as a function of the emergence of modern scientific practice. So much can be readily (if summarily) demonstrated by juxtaposing a statement from Nicholas Copernicus's *De Revolutionibus Orbium Coelestium* with one from Isaac Newton's *Principia Mathematica*. In 1543, Copernicus's text—canonized, of course, as a foundation for modern scientific rationalism—nevertheless qualified the status of its heliocentric assertions: "let no one expect from astronomy, as far as hypotheses are concerned, anything certain, since it cannot produce any such thing."[52] Some 120 years later, in 1686, Newton's summary General Scholium posited a contrary position: "Hypotheses non fingo"—"I do not feign hypotheses."[53] Obviously, something had happened to the status of hypotheses in emergent science: where the earlier text seems to have found solace in an assertion of their provisionality, their oblique promise of uncertainty, Newton's *Principia* takes pains to insist that any imputation of "feigning" has no place in his usage. This change in status is one index of an important cultural shift, out of which modern taxonomies of writing begin to produce and reproduce themselves. In effect, the hypothesis had been reinscribed in a new discursive formation: that of "science" as opposed to "fiction," the poetic.

That the hypothesis was a flash point for the truth claims of emergent scientific rationality is suggested by the word's mobility in classical Greek writing. The term "hypothesis" had no single domain, having been employed in rhetorical, geometrical, and poetical contexts to represent, broadly speaking, what must be stipulated as a formal starting point for further demonstration or staging.[54] Perhaps most immediately significant to Renaissance debates was Aristotle: the *Poetics* was of course influential in late humanist debates on the epistemological status of poetry, and in it he used the term to designate the *argumentum,* or plot, of a drama.[55] Indeed, humanist interest accrued around Aristotle's text as the site from which a discourse of legitimation for the study of "letters," as vocation and as emergent profession, could be launched. Under its sign, both poets and critics offered pronouncements about genre, utility, and mimetic function. For instance, both Julius Caesar Scaliger and Lodovico Castelvetro wrote that poets could claim the right to imitate truth by means of fiction and thereby circumscribe the binding of representation to mere facticity.[56] It is not far from these contextual accounts to Sir Philip Sidney's absolutism. A humanist poet who "ranges . . . within the zodiac of his own wit" may only be laying metaphorical claim to the objects of astronomy; nevertheless, his insistence that such a poet is not tied to nature in his framing of formal propositions about the world—unlike the astronomer, the geometrician, or the rhetorician, among others—signifies that domains of expertise once covered by classical usage of the term *hypothesis* are being contested.[57]

The humanist legitimation of poetry affords a way to see the consolidation of scientific ideology from its obverse—which is to say from a differing and distinguishing relationship to truth-claims.[58] It is therefore conceivable that the regimes in which scientific discourse increasingly invested were de facto problematized by the term.[59] Witness the flurry of contemporary commentary about the status of hypotheses in *De Revolutionibus,* made prominent through the "Address to the Reader" that prefaces Copernicus's more technical treatise. Martin Luther believed Copernicus had dared too much, while for the dialectician Petrus Ramus, he had proven too little: "The fiction of hypotheses is absurd . . . would that Copernicus had rather put his mind to the establishing of an astronomy without hypotheses."[60] It seems that few might have agreed with the position expressed by Nicolas Reymer Baer in 1597, that hypotheses, as "fictitious descriptions of an imaginary form of the system of the world" would "not be hypotheses if they were true."[61] Johannes Kepler, for one, mocked Reymer Baer's error in equating the hypothetical with the fictional.[62] Kepler had in fact revealed that the address had been written by Andreas Osiander—a Protestant cleric who chose not to sign his name—rather than by Copernicus himself, and that its stance toward hypotheses did not reproduce the practice in the rest of the text. Thus, he wrote that "the most absurd fiction . . . that the phenomena of nature can be demonstrated from false causes" was not to be found in Copernicus, who "thought his hypotheses were true" and was not "composing a myth but is giving earnest expression to paradoxes, that is, philosophizing"—unlike Osiander, "the author of this fiction."[63]

However much Kepler's desire to preserve the original intent of the treatise's author has been duplicated by subsequent historians, from the standpoint of emergent disciplinary boundaries the address's engagement with the author-function does not seem significant. Which is not to debar scientific treatises from the function, as the case of Galileo, the hero of modern science, shall suggest. Here, however, who is responsible for the address is less important than the fact of its existence: it provided a frame for understanding how the propositions in *De Revolutionibus* were read and understood in the sixteenth and seventeenth centuries. The inconsistency Kepler marks out between Osiander's denial of truth-status to the hypothesis and the demonstrations that depend upon the contrary occurs within a general cultural problematic, where the knowledge-claims of humanist-inflected practice collide with a growing impetus toward the disembodied (Cartesian?) certainties of methodical abstraction.

The disciplinary boundaries whose pre-formation I am considering have all too seldom been suspended in historiographical practice focusing on the emergence of science as an object of study. Yet as Robert

Westman has persuasively demonstrated, it is possible to approach *De Revolutionibus* from the "other side," as it were, which is to say from the standpoint of its relation to humanist modes of textuality.[64] By examining a copy of Copernicus's treatise annotated by the sixteenth-century scholar Michael Maestlin, Westman discovers an allusion in the "preface" to Horace's *Ars poetica* and its doctrine of decorum that leads him, in passing, to consider the aesthetic ramifications of Copernican rhetoric. More may be made of the connection, as I shall suggest. It is not solely that Copernicus's preface—more properly, the dedicatory epistle to Pope Paul III—is revealed as suspended in a web of intertextual and interdiscursive reference clarified through humanist annotation. It is that the epistle itself maps out the condition of its own discursive intelligibility in such a way as to render problematic the ready assignation of the treatise to the history of science.

Unlike the address to the reader that has occasioned authorship debate, the dedicatory epistle is signed by Copernicus. However, it may be juxtaposed to that vexed piece of writing, since it manifests a similar confusion, whether strategic or not, about the sort of adjudication to which the text will be subjected. If read preliminarily as essays in authorial sincerity, both the address and the epistle seem to map out a transparent anxiety about opposing the scholasticism of church astronomers. Osiander's address, for instance, alludes to the heliocentric hypothesis in a syntactical aside: the main force of the period rather falls upon an anticipated objection to its content:

I have no doubt that certain learned men, now that the novelty of the hypotheses in this work has been widely reported—for it establishes that the Earth moves, and indeed that the Sun is motionless in the middle of the universe—are extremely shocked, and think that the scholarly disciplines, rightly established once and for all, should not be upset.

[Non dubito, quin eruditi quidam, vulgata iam de novitate hypotheseon huius operis fama, quod terram mobilem, solem vero in medio universi immobilem constituit, vehementer sint offensi, putentque disciplinas liberales recte iam olim constitutas turbare non oportere.][65]

The "author of this work" must not be regarded as a transgressor against accepted astronomical practice, since he has "committed nothing which deserves censure." Rather, the text presents him as a collector of "diligent and skilful observations," who beholds and then constructs "laws . . . or rather hypotheses." Here the distinction between laws and hypotheses seems locally strategic: since "true laws cannot be reached by the use of reason" (22), it is preferable to draw on hypotheses because it is unnecessary as well as impossible for them to be "true, nor indeed even probable." Thus "*novas hypotheses*" may take their place beside the alternative cosmological schemes that they will come to supplant (at least in the "Vulgar Triumphalist" model of historical change):[66]

Let us therefore allow these new hypotheses also to become known beside the older, which are no more probable, especially since they are remarkable and easy; and let them bring with them the vast treasury of highly learned observations. And let no one expect from astronomy, as far as hypotheses are concerned, anything certain, since it cannot produce any such thing, in case if he seizes on things constructed for any other purpose as true, he departs from this discipline more foolish than he came to it. Farewell. (22–23)

[Sinamus igitur et has novas hypotheses inter veteres nihilo verisimiliores innotescere, praesertim cum admirabiles simul et faciles sint, ingentemque thesaurum doctissimarum observationum secum advehant. Neque quisquam, quod ad hypotheses attinet, quicquam certi ab astronomia expectet, cum ipsa nihil tale praestare queat, ne si in alium usum conficta pro veris arripiat, stultior ab hac disciplina discedat quam accesserit. Vale.] (30–31)

"Vulgar Triumphalism," however, seems to be afforded little purchase on the care manifested in the address—and later, too, in the epistle to Pope Paul III—in circumscribing reception and opinion. Osiander's demand for laissez-faire around the hypothesis as well as his sense of embattlement surely confirm that it is important to place the address in context. A politic negotiation of the scholarly circuit, and not necessarily cowardly epistemology or foolishness, appears to dictate in the first instance how strong the truth-claims of the Copernican hypothesis are seen to be.

Yet such a local account could itself be turned around by attention to the wider ramifications of novelty in the sixteenth century. When Osiander flags the hypotheses as *"novas,"* he also renders them susceptible to the wholesale critique to which Donne, for instance, subjects formations of the new. The sixteenth-century contestations around the hypothesis that I have cited are in fact versions of such a critique: although they seem to insist that hypotheses be *more* scientific, more "new," it is the fact of attention itself that seems novel—that seems, in fact, part of the general semiosis of modernity.

It is in light of this semiosis that Copernicus's dedicatory epistle should be examined. The milieu out of which it emerges and to which it aims to speak is conservative, backward-looking; it is also text-bound. Even if Aristotle rather than Cicero is the tutelary deity, scholastic astronomy shares with humanist philology the authority of the text as a conduit to classical presence. If, then, Copernicus's "anxiety" is reexamined, it seems to disperse into a network of allusions and stances typical, I am tempted to say, of humanist rhetoric. When Copernicus imagines that there will be a "clamor for [him] to be hooted off the stage," the terms themselves suggest a play around the dramatic, and familiar, sense of the term "hypothesis":

I can well appreciate, Holy Father, that as soon as certain people realize that in these books which I have written about the Revolutions of the spheres of the universe I attribute certain motions to the globe of the Earth, they will at once *clamor for me to be hooted off the stage* with such an opinion. (23; emphasis added)

[Satis equidem, Sanctissime Pater, aestimare possum, futurum esse, ut, simul atque quidam acceperint, me hisce meis libris, quos de revolutionibus sphaerarum mundi scripsi, terrae globo tribuere quosdam motus, statim *me explodendum* cum tali opinione *clamitent*. Neque enim ita mihi mea placent, ut non perpendam, quid alii de illis iudicaturi sint.] (35; emphasis added)

Albeit in passing, Copernicus's language offers an image of *De Revolutionibus* as a spectacle, a form of entertainment, with the author figuring most prominently as an actor—one who fears his hypotheses will bring forth a laughable performance on the stage of public opinion:

[S]ince I was thinking to myself what an absurd *piece of play-acting* it would be reckoned, by those who knew that the judgments of many centuries had reinforced the opinion that the Earth is placed motionless in the middle of the heaven, as though at its center, if I on the contrary asserted that the Earth moves, I hesitated for a long time whether to bring my treatise, written to demonstrate its motion, into the light of day. . . . (23–24; emphasis added)

[Itaque cum mecum ipse cogitarem, quam absurdum ἀκρόαμα (*akroama*) existimaturi essent illi, qui multorum seculorum judiciis hanc opinionem confirmatam norunt, quod terra immobilis in medio coeli tamquam centrum illius posita sit, si ego contra assererem terram moveri, diu mecum haesi, an meos commentarios in eius motus demonstrationem conscriptos in lucem darem.] (36–37; emphasis added)

The Greek word "*akroama*" here, like *explodendum* in the preceding passage, offers a context for Copernican hypotheses that seems pointedly—atavistically?—dramatic. It calls attention to the constructed nature of the theory being advanced, and it insists, as well, on the pertinence of a social milieu represented as a site of aesthetic adjudication. In the passages cited, the concept of "hypothesis" turns subtly in upon itself; the "new" hypothesis, of assertion about the natural world, converges upon the Aristotelian hypothesis of text and stage. Conviction thus depends upon effective performance, effectively scripted. Like the actor who is only as convincing as the material he is given to recite, Copernicus's epistle figures him at the mercy of an audience well-versed in Aristotle and skeptical about the plausible impossibilities of his theory.

Yet as the context suggests, those standards are aesthetic, and their force a matter of taste, convention, opinion, rather than a strict and "scientific" rectitude. To risk overstatement: if Copernicus's hypotheses are plots, then their critics cannot deem them true or false, but only good or bad. The possibility of reading the text in this way holds, whether the play in language is a matter of intent or the fallout of discursive differentiation: they are the same thing in effect. Given this correspondence, Kepler's outrage against Osiander for having forced "a most absurd fiction" on *De Revolutionibus* shows, among other things, the uneven status of that differentiation in the sixteenth and seventeenth centuries: if for Kepler true hypotheses could easily be framed and distinguished from other discursive instances, the address and the epistle, when taken together, suggest the converse, suggest a dispensation toward the natural world whose instruments of inquiry are still very much within the purview of humanism, or text-driven scholasticism. The reference to Horace's *Ars poetica* also suggests as much: the letter to the pope seems caught between aesthetic categories, like the decorous or the pleasing, and nascently scientific ones; between a humanist notion of textual authority and immediacy and a prolepsis of facticity.

To read the ancillary material around *De Revolutionibus* as I have done throws into relief precisely the form of novelty, discursive innovation, that can be attached to Copernicus's controversial treatise. Even more, however, it exemplifies how the texts of the sixteenth and seventeenth centuries themselves, as sites inscribed by contradictory discursive positions, serve as archives for the isolation of scientific modernity they also incubate.

To call them archival, however, is not to deny those counterinscriptions their agency as discourses. Thus I want to turn to the issues with which I commenced: the way in which the emergent scientific ideology signified by such labels as the "New Science" or the "Copernican Revolution" processes itself through notions of otherness, and the consequences of that transumptive move for "literature" as a discursive production within history. A tropological connection between the New World and the New Science was significant, in the first instance, in providing the emergent field of inquiry with sufficient cultural justification for its own importance. When, for example, Thomas Harriot—a notable mathematician as well as an agent of the Virginia Company—receives a letter that says "Galileus hath done more in his . . . discoverie [with the telescope] than Magellane in openinge the streightes to the South sea," the connection exceeds a recognition, on the level of writing, that the superior information made possible by more accurate astronomical work enabled the circumnavigation of the globe.[67] Such persistent coding of the emergent investigation of nature through an exploratory matrix imports into the coding text the symbolic importance of discovery as it attaches to the names of voyagers; in addition, it tacitly posits the future of the New Science as a parallel to the history of colonization.

This rhetoric of affiliation, in turn, installs the enterprises of canonical protomodernists like Copernicus—or Galileo, or Bacon—in a political history, and aligns them with the conditions from which, and partly because of which, they emerged into discursive power. In this way it is possible to reconfigure the Baconian connection between his programmatic attempts at a geography of knowledge and Columbus's belief in the existence of unknown worlds in the face of arguments to the contrary. Formerly a sign of heroic individualism, the connection can now be seen as a forecast of what it suppresses in representation: discovery's successor, colonization, the discourse of the other that is always a discourse of mastery. In such moments, one agent of early modernity legitimates another.

Michel de Certeau asserts that the structure of modern Western culture is heterological, which is to say a discourse of the other. The *"intelligibility"* of this modern culture *"is established through a relation with the other; it moves (or 'progresses') by changing what it makes"* of its other.[68] Given this formulation, what are the conditions for intelligibility in the emergent scientific ideology I have been invoking throughout? Another way to put this is to ask what scientific "discovery" does with the terrain it attempts to transform to its own desires. The modernity with which I have been concerned demands a past against which to reinvent itself: thus the body of the Algonkian, which expresses itself through its physicality, and hence is metonymous for the object that cannot be heard to speak itself, identify itself, is one place to begin. But the answer does not rest with reinscribing the conditions of brute appropriation, with talking about science's domination and "the death of Nature" purely through a nostalgic discussion of the natural body. What access is there to the early modern body, whether human or metaphorical, when it is not rendered textual?

The conditions of intelligibility for a scientific ideology in the process of emergence are therefore discursive, foregrounding the issue of textuality, both because "nature" is an ideological construct that exists in language, and because the figures who inaugurate scientific modernism must be considered as propagandists—agents, whether intentional or not, of a new discursive formation. So much is suggested by the "*novas hypotheses*" of Osiander and Copernicus. Here, too, the "innovations" of Ignatius Loyola, Donne's keeper of hell-gate, assume their symbolic importance beyond the text that already posits the Jesuits as New World colonizers: the word "propaganda" comes from a Jesuit institution, founded in 1622, for the propagation of the faith. An organ of the Counter-Reformation, the *Congregatio de Propaganda Fidei* was also an agent of European hegemony in the non-European world, a protomodern bureaucracy for which the mission of converting the Other was coextensive with the act of learning his language.[69] As the example of the Jesuits suggests, the modern becomes intelligible and achieves its conquest through inhabiting, ultimately ventriloquizing, alien forms of signification.

What this move then highlights is representation as the contested ground, if by representation is understood the humanist tradition of textuality, affiliated with a past now available as such, and posited as the inaugural term in a narrative of development. The change from a past discursivity, ideologized as transparent and immediate, to a past posited as the raw material of the present enables the reemployment of humanist forms of representation by emergent narratives of science. More's *Utopia,* for example, becomes a prime site for the negotiation of difference and emergence: it is as though More's own instantiation of the New World in his text—although a New World whose own internal text is the Plato of classical antiquity—makes clear the colonialist imperative of emergent science. This exemplary instantiation, in turn, makes the text peculiarly appropriate for discursive reconfiguration— as it is by Bacon in *The New Atlantis,* and less directly by Galileo through the thought-experiments he uses in his Copernican dialogue. Bacon, for instance, rewrites the *Utopia* in his *New Atlantis* and turns

More's textual equipoise into a mechanism for the generation of the modern subject of science. Galileo, in his experiments in immateriality (what are now called thought-experiments), completely empties out utopian form of humanist content in what might be a limit-case of discursive colonialism: the thought-experiment becomes a fantasia—hypothesis—of ideality, of conditions free from ideological constraint that are themselves the ideology of science.

Let me circle back to where these arguments began. The presence in Donne's hell both of Columbus and Copernicus suggests their equivalence as avatars of the new, but it also argues the utility of one in giving form to the operations of the other. Humanist literary narratives become a way to give a metaphorical body to the emergent narrative of disembodiment, to colonize the form and to reconfigure the content. Like the emblematic moment of disembodiment in Descartes's cogito, where the body of the observer is turned by linguistic operation into an object for the scrutinized subject of pure consciousness, the emergence into discourse of scientific ideology, through its affiliation with America, acknowledges its fantasy of a free space—but one informed always by what it excludes, processes, renders unproductive.[70]

More than formalist issues are at stake in such contestations, hinging as they do on the assignation of productive value in a culture within which, if Donne's text is any indication, innovation—the making of new things, new forms—is on the rise. The novelty whose production I have been excavating must be understood as necessarily imbricated within other cultural manufactures, whose presence I have only alluded to—the tying of gender to the evidentiary basis of biology, as in Andreas Vesalius and his successors; the increasing dissemination of technological prostheses, like telescopes and microscopes, which stand in for the body of the European observer removed from the scene of action. These would take me into the realm of a materiality that must be invoked to offset the pure ideation of novelty that I have generally been concerned with here—the material body that, as *The Tempest* indicates, works both for science and colonialism, for Prospero the magus as well as Prospero the European settler.

2

Admiring Miranda and
Enslaving Nature

It matters to know how sex and nature become natural-technical objects of knowledge, as much as it matters to explain their doubles, gender and culture. It is not the case that no story could be told without these dualisms or that they are part of the structure of the mind or of language. For one thing, alternative stories within primatology exist. But these binarisms have been especially *productive* and especially *problematic* for constructions of female and race-marked bodies; it is crucial to see how these binarisms may be deconstructed and maybe redeployed.

—Donna Haraway, *Primate Visions*[1]

In a sense, Donna Haraway's influential explorations of the interrelationships among gender, race, imperialism, and the scientific in *Primate Visions* provide the intertexts for my reading of *The Tempest*. Not that the argument of this chapter depends on a teleological narrative, nor is it the case that Haraway has revealed something essentially stable about the structure of scientific modernity. Rather, the isomorphism between the colonial and the scientific—what in this study has been flagged through tropes of novelty as the "New World" and the "New Science"—reveals the historical coincidence of two modes of power-knowledge, of conquest, at their emergence. These modes, moreover, are often mutually constitutive, interdependent, given their intermittent rearticulation within successive cultural formations. Through its close reading of historical artifacts, Haraway's study of the conduct and assumptions of modern-day primatology manifests the interlacing of Western imperialism and Western knowledge-seeking agendas. In both cases, the terms of relation are constituted by the "marked bodies of race, class, and sex," which have been "at the center, not the margins, of knowledge in modern conditions. These bodies are made to speak because a great deal depends on their active management" (289). Science and imperialism, then, are essays in resource control as well as political enterprises. Their designated objects are deployed, in extreme practices enslaved, in part to serve particular ideological and material ends, but also, in part, to attest to discursive power through materiality, as chapter 1 has been at pains to suggest. This inadvertent signification—the "speech" of interrogated, subjoined bodies—is the location from which dominant inscriptions of "man," and of "nature," scientific as well as humanist, have emerged.[2]

Haraway's work is continuous with overt critiques of the agendas legitimated and naturalized in the name of science. To cite but one example, Sander Gilman has analyzed the horrific exploitation to which the "Hottentot Venus"—a nineteenth-century African woman named Saartje Bartmann, who was deemed to possess freakishly large genitals—was subjected.[3] After being brought to Europe, Bartmann was put on public display, to mixed outrage, titillation, and scandal; she died about five years after, when her "organs of generation" were given over to pathology research. The scientist Georges Cuvier's postmortem clinical fascination with her genitalia seems an obscene parody of what is ideologically processed as dispassionate scientific observation; instead it resembles a cultural deformation, where visual interrogation and exploration are weirdly energized by their predication of a disenfranchised African female as their object of inquiry. But of course it is not clearly any such thing; rather, as Haraway's complementary and corrective account indicates, the transmogrification of Saartje Bartmann into the "Hottentot Venus" shows the historical dependence of the scientific on prior ideological investments—investments concerning gender and race, among many other possibilities, which the authority of the scientific can subsequently be brought in to ratify as objective truth.[4]

Gilman observes that Buffon and others connected the sexual appetites of Africans with those deemed to be their near relations: the apes. Haraway is particularly interested in what science makes of primates because they "have a privileged relation to nature and culture for western people: simians occupy the border zones between those potent mythic zones" (1). In current scientific practice, apes, not unlike the subjects of ethnographic fieldwork in nonindustrial cultures, constitute test cases for the elasticity, the potential inclusiveness, of the "family of man": so Haraway's exemplary case studies indicate. In order both to manufacture and contest the border zones that juxtapose humans with their doubles in "nature," much work in the field of primatology has tried out models of human signification upon a population deemed, in potential at least, to be almost human—as when, for instance, the gorilla Koko is instructed in American Sign Language and encouraged to keep a pet.[5]

But to first pose a question that will reappear later on in a related context: does it matter who teaches languages to the apes? This question intends to get at a specific aspect of Haraway's work, and so to set up my analysis of the early modern structures analogous to those she discusses in connection with the postcolonial and global production of science. What is the particular place of women in primatology, as subjects of knowledge, as agents of social reproduction? Such issues are sited at the heart of power-knowledge, as Haraway's words exemplify:

"the politics of being female" are at the origin of western order, including scientific accounts of what it means to be human, to be female, and to be an organism. But women's authorship of those politics is not, literally, "original." . . . In the west—including western science in all its foundational mythic moments of origin with "the Greeks" or at the great instauration of "the scientific revolution" or at the moment of Darwin's transformative account of "the origin of species"—to be female has been to be a pretext, not an author and a subject of history. (280)

This constitutive belatedness poses a dilemma (often unacknowledged) for women who participate in research agendas like those taking place within modern primatology and, indeed, in all forms of knowledge that Haraway designates as "bio-political." Given that the organism or the family, or some definitional substratum of the human or material, is at the basis of all these knowledge-seeking practices, women have increasingly been allowed into their purview—not so much as male-identified exceptions but almost as native informants.

Hence, within Haraway's account, women's research and fieldwork positions the feminine as intercessory, even liminal: by shaping fields of investigation and by training alien subjectivities in access to the dominant, hominid model of signification, they reproduce themselves as the portals of and to nature, while standing, it seems, more firmly in culture as a result. Haraway's female primatologists are crucial to the work of social and ideological reproduction embedded in the specific practices of twentieth-century scientific investigation. As part of a symbolic repertory, they are crucial, too, as indirect registers to the systematic transformation and predication of nature in scientific modernity. Yet because women have historically constituted the marked (which is to say embodied) gender, they are also a part of what gets differentially suppressed in the new discursive formations that science brings into prominence. Haraway's adroit discussion makes it possible to compare such important primatologists as Jane Goodall and Sarah Blaffer Hrdy to the figural Lucretia of early humanist tradition, or the unsignified mother who does not participate in Petrarch's acquisition of language. Although these scientists clearly possess more instrumentality than the mythic Lucretia—after all, they are real women—like her they are vehicles, vessels of transmission, cultural conduits. If these women are allowed substantially more agency in the later days of modernity, it is because the divisions in the epistemological field necessitate the ever more precisely ergonomic assignment of life sciences to living women.

This is where *The Tempest*'s easily forgotten Miranda, Caliban's first tutor in language, fits in. Her intermediary role in the discursive production that he represents sets up her complicity in defining the border zone between nature and culture in the colonial fantasy at the moment of its earliest coalescence. When the civilizing process fails—or perhaps, as the analysis later in this chapter suggests, when Miranda succeeds too well at it—Caliban is all but discarded as a failed experiment, a creature "on whose nature/Nurture can never stick."[6] The productive ambiguity over Caliban's position made possible by Miranda's interventions with him place the "salvage and deformed slave" at a discursive crossroads—one way leads to early modern ethnography, the other, however distantly, to primatology.

Haraway's analysis, however, depends on the prior availability of "science" as an accomplished discursive category within which another subfield or subcategory can be inscribed. During the early modern period the boundaries of such discourse were being consolidated in important ways; witness the competing, even parallel epistemological claims raised by the occult tradition.[7] There has been much historiographic work done to distinguish, or to refuse to distinguish, the legitimated knowledges that count as science from their "magical," delegitimated counterparts. For instance, one attempt to comprehend the apparent strangeness of Renaissance occultism draws on anthropological modeling and seeks to compare it to other "primitive" systems.[8] As so-called traditional epistemologies, Renaissance magic and the belief systems of, say, African tribal cultures, lack a sense of an "outside" to their knowledge, of alternative theoretical conceptualizations that are acknowledged as such, and that can be resorted to when the traditional beliefs are under siege. Both, therefore, are "closed," in comparison to the "openness" of modern, "scientifically-oriented cultures."[9] The antithesis between open and closed, primitive (or traditional) and modern, is particularly productive for a scientific ideology imbricated within colonialist agendas, as Haraway's work certainly evinces. And in chapter 3 I will argue for the ideological force of that openness as it forms, determines, Francis Bacon's "unperfected" colonial-utopian text, *The New Atlantis,* and the scientific subject who reads it. Here, however, I want to consider the symptomatic function of the concept of the primitive, filtered through the autocritiques of postcolonial ethnography.[10] As Johannes Fabian, James Clifford, and others argue, European fieldworkers have called primitive those cultures—and by extension those knowledges—that have been subjugated or objectified in the wake of Western imperialism and the regimes of truth with which it is accompanied. This subjugation is not always and immediately in material fact; in the case of traditional knowledges, for instance, it is the work of a taxonomy that validates the superior "openness," and systemic resiliency in the face of falsification, of European epistemological claims. Thus occultism may indeed be "primitive," but only in the interested sense given to that term within dominant ideologies of rationality. As with the African tribal cultures to which it is compared, the early modern occult may interest researchers precisely as a function of its exoticism and difference—and thus through the discursive effects of the colonizing agents of modernity.

The status of Renaissance magic as an epistemology is related to the tropes by which cultural change is narrated. How can we recognize the constellation of race, gender, colonization, and nature—key concepts in Haraway—as these concepts emerge, in a text that reveals the conditions of their preformation in, and *as,* discourse? Such concepts, although undoubtedly forms of relation useful to the study of the Renaissance, are also the product of Enlightenment taxonomies, of the "critical rationalistic mentality" that Brian Vickers has defined against the symbolic and metaphorical reasoning of occultism.[11] In examining the practices concerning natural forces that are now distinguished from "scientific" ones, Vickers formulates a model of occult reasoning that emphasizes the instrumentality of signification: "The occult sciences' practice of substitution or interchangeability of concepts depends fundamentally on the reification process, the breakdown of the line between literal and figurative" (122).

But the more important question is whether occult texts manifest the breakdown of such a line, or rather its uneven manufacture as a condition of a nascent scientific modernity. In other words, what is the relationship between occultism, science, and the power of the word, as over against the newfound reality of the thing, the object-world?

Given an unexpected sympathy between Vickers's analysis of occult analogy and the archaeology of Renaissance correspondences produced by Foucault in *The Order of Things,* it may be that we are to look in "primitive" forms such as literature for evidence—signs—of the preformation of colonialist science and of its relation to simultaneous reworkings of race and gender.[12] In his exploration of the human sciences Foucault has provocatively suggested that the Renaissance episteme of analogy, resemblance, and correspondence, although generally supplanted by later models of ordering and reason, may still be among us in a peculiar, or perhaps merely unacknowledged, throwback:

There is nothing now, either in our knowledge or in our reflection, that still recalls the memory of that being. Nothing, except perhaps literature—and even then in a fashion more allusive and diagonal than direct. It may be said in a sense that "literature," as it was constituted and so designated on the threshold of the modern age, manifests, at a time when it was least expected, the reappearance, of the living being of language. In the seventeenth and eighteenth centuries, the peculiar existence and ancient solidity of language as a thing inscribed in the fabric of the world were dissolved in the functioning of representation. . . . The art of language was a way of "making a sign"—of simultaneously signifying something and arranging signs around that thing; an art of naming, therefore, and then . . . of capturing that name, of enclosing and concealing it, of designating it in turn by other names that were the deferred presence of the first name, its secondary sign, its figuration, its rhetorical panoply. And yet, throughout the nineteenth century, and right up to our own day . . . literature achieved an autonomous existence . . . only by forming a sort of "counter-discourse," and by finding its way back from the representative or signifying function of language to this raw being that had been forgotten since the sixteenth century.[13]

Foucault's speculations on the literary as survivalist discourse should be invoked with caution. Most problematic, perhaps, is his invocation of the Renaissance as a site both of immediacy and of loss, a moment before representation when language had "living" or "raw being." His rhetoric is not incidentally primitivist or Edenic; although less typical of his later writing, its nostalgia may explain why Foucault seems never to have taken up Renaissance epistemes again, except as a formal point of departure.[14] Nevertheless, something may be made of the zone of difference, of exemption from the dominant relations

among post-Enlightenment disciplines, that Foucault accords to literature in this early study; it serves as a useful (even if provisional) way to
account for the gradual deterioration of its epistemic authority in a
subsequent regime, a subsequent modeling of "truth." If we grant literature the mythic place Foucault hypothesizes for it, the texts we now
call literary become relics of analogical relation, and akin to the occult
practices described in Vickers's formulation. It stands to reason, therefore, that to examine the correspondence between New World and
New Science—a connection whose later relationship is productive for
Haraway, but one that at first glance seems merely analogical—we
should search for their traces in literary texts.

Hence the role of *The Tempest* in this study. Of course, it is highly
pertinent that the play has, until recently, been much read as a magical
fiction—and that the reading which by and large has supplanted it is
colonialist.[15] The sign of that fiction, and of magic both, is Prospero's
books. They are the source of Prospero's knowledge and authority, basis
of his control over the spirits of the island and over Caliban, its more
material inhabitant, and ultimately, too, over Miranda, his daughter.

Chapter 1 examined how the Algonkian of Harriot's *Report* was
subjected to the protoethnographic gaze and suggested, in consequence, that the New World inhabitant signified through his embodiment. More will be said about New World inhabitants as instantiations
of the natural, especially in light of the now-dominant interpretation of
The Tempest as a colonialist allegory. But the transformation of alien
bodies into natural objects cannot be effected without an examination
of what predicates the transformation, what positions those bodies as
acted upon. This is not simply an issue that devolves on Prospero's
authority; it also depends on the site of that authority, the discursive status of his knowledge—the fact that it is contained in books,
in objects full of words. In examining the priority granted language
within occult practices, Vickers suggests how words are predicates of
phenomena: "Words are treated as if they are equivalent to things and
can be substituted for them. Manipulate the one and you manipulate
the other" (95). The commutativity between nature and significa-

tion, with the practical consequence that magical practices need books, broadly connects occultism with humanism; more specifically, magic becomes congruent with a form of protoscience that has been deemed humanist.[16]

Notably, Prospero proclaims himself an adept in the liberal arts, as much a humanist as a magus. Moreover, the play is much concerned to reiterate that his books are seen, by one key observer at least, to be the locus of power. Witness Caliban's repeated remonstrance to Trinculo and Stephano:

> There thou mayst brain him,
> Having first seized his books. . . .
> . . . Remember
> First to possess his books; for without them
> He's but a sot, as I am, nor hath not
> One spirit to command—they all do hate him
> As rootedly as I. Burn but his books.
> (3.2.86–87; 89–93)

Appropriately enough for a colonialist allegory, the power of the word in this instance cannot be entirely extricated from its status as an instrument of subjugation, as a form of control almost magical in itself; so Lévi-Strauss in his well-known anecdote of the writing lesson in *Tristes Tropiques* makes clear.[17] The chief of the Nambikwara whom Lévi-Strauss observes and writes about simulates the ethnographer's note-taking, borrowing writing to serve "as a symbol, and for a sociological rather than an intellectual purpose. . . . It had not been a question of acquiring knowledge, of remembering and understanding, but rather of increasing the authority and prestige of one individual . . . at the expense of others" (298). Indeed, writing endows literacy with a certain compulsoriness as well as status; as a vehicle of ideological production that Louis Althusser would recognize, it extends the dominion of authority irrespective of physical force and makes slavery and empire possible (299–300).

Obviously, then, the fact that Caliban fetishizes Prospero's access to written (or printed) signification need not be separated from its power to do magic. Quite the contrary. If magic "fails" historically as a discourse, it is in taking on a task for which it is not suited. We may almost say that the impossibility of magic, its historical untenability, lies in its attempt to bring together a written legacy with a desire for instrumentality over the phenomenal world. By aiming both at socio-discursive and material control, magic becomes an attempt to occupy two increasingly different ideological registers. In this regard, Prospero's "rough magic," extinguished when he drowns his book, evinces its compatibility with the text-driven culture of humanism at the same time that it points to the limitations of both in the production of modernity.

The Tempest clearly reveals those limitations, to show the contingency of the books deemed elsewhere in the play-text to be all-powerful. For one, Prospero's knowledge appears to be useless in the Dukedom of Milan, stands opposed, in fact, to the power of the throne he is born to occupy, as the *vita contemplativa* to the *vita activa*.[18] The "library [that] Was dukedom enough" (1.2.109–110) is an inadequate, because immaterial, substitution of the *res publica litterarum* for the "temporal royalties" of Prospero's civil power. For another, even on the island, the books seem initially to have had little pragmatic value: it is Caliban who must show Prospero and Miranda the sources of fresh water and edible fruits, to enable them to survive on unfamiliar soil. In this regard, at least, Caliban has the early advantage in the game of power-knowledge to be played between them.

Yet, indisputably, something critical, something instrumental, happens with and to the "secret studies" (1.2.77) by which Prospero has been rapt and transported: the inaugural storm is sign enough of the change. The difference in efficacy between Prospero's books in Milan and in exile is provided by the island itself, which quickens—activates—Prospero's power. The conjunction of his dominion over the forces of nature and his political domination of a colonial setting underscores the isomorphism of the "New World" and the "New Science" that I have insisted upon. It may well be objected, however, that I have already claimed Prospero as a magus: how then to transfer him to the realm of the nascently scientific? Again, the answer lies in part with the island, its status as utopian space, a space for the readier configuration of new social and ideological relations. In that sense, *The Tempest* sketches a problematic that comes to fulfillment with Bacon's *New Atlantis:* the utopian island precipitates a fantasy, not of the library, but of the laboratory. In the case of Prospero, however, it is as though the colonialist aspects of location itself enable book-bound magic to be turned into a species of science. Unshored, however forcibly, from his European context, Prospero becomes as a consequence a version of the Icarian subject described by Carlo Ginzburg. Not that Prospero's deeds themselves resemble the experimental agendas of Bacon, Harriot, or Galileo; on the contrary, his summary recitation of his accomplishments aligns him firmly with witchcraft and black arts, "a world of Medeas, of Thessalian witches":[19]

> Ye elves of hills, brooks, standing lakes and groves,
> And ye that on the sands with printless foot
> Do chase the ebbing Neptune, and do fly him
> When he comes back; you demi-puppets . . .
> . . . And you whose pastime
> Is to make midnight mushrooms, that rejoice
> To hear the solemn curfew, by whose aid—
> Weak masters though ye be—I have bedimmed
> The noontide sun, called forth the mutinous winds,
> And 'twixt the green sea and the azured vault
> Set roaring war . . . (5.1.33–44)

Note the tropes of political divisiveness that recur through his speech, in the "mutinous winds" and "roaring war" between the sea and sky. This transference of civil language to the elements may simply underscore the analogical basis of magic and conjuration both, of macrocosmic and microcosmic correlation as noted by an agent with some stake in ratifying the connection between the natural order and the political one: otherwise, there would be no point in the storm and what it sets in motion. Yet before Prospero admits to such strong (and Sycorax-like) activities as raising the dead, he invokes the "weak masters" who have been his agents. Despite his subsequent reversion to the first person, Prospero needs operants in the phenomenal world, extensions of himself beyond his reading and signifying mind. Of course, Ariel and his cohorts are not neutral forms of instrumentation, mere devices for the performance of Prospero's sovereign will; it takes science proper to supplement the inquirer into nature with a technological armature. But the capacity for self-extension with which the island provides Prospero—even if couched in animistic terms—begins, remotely, to figure the productiveness of colonial space in the making of modern epistemologies.

This shading-over of occult writing into colonialist science in *The Tempest* leads back to the distinction that the play effects between the regime of the signifier and that of the material agent. Words, as Caliban should perhaps have remembered from his own lessons in language, do not seem to make very much happen: one needs "spirits to command" as well as the books that tell one how to command them. Or, rather, what words do make happen is that form of control that I have already deemed social and discursive. When Caliban is taught to name the greater light and the less, he is not getting a lesson in astronomy so much as he is being interpellated in the hierarchical construction and disposition of the cosmos. Without forces at his disposal, or colonial subjects over whom to exert dominion, Caliban can have no use for the finer points of knowledge that appear—only appear—to be strictly bound between the covers. After all, Miranda presumably has had access to these texts all along. And yet she, not much less than Caliban or Ariel, is positioned outside their ken.

In an essay published in the 1980s, Jacqueline Rose has discussed the "fantasy of the woman" as a symptom of reading with respect to *Hamlet* and *Measure for Measure*.[20] How far, she asks, "has the woman been at the centre, not only of the internal drama" (by which she means the Shakespearean play-text as represented) "but also of the critical drama—the controversy about meaning and language—which each of these plays has provoked?" (95). By way of answering these "accusations," she examines the modern critical tradition that fetishizes women's sexuality both as site of interpretation and as symptom of the problematic hermeneutics of each play. Whether discussing the legendary Oedipality of *Hamlet* or the "obsessive," "hysterical," or "saintly" Isabella, male critics have longed to fix their readings (and their difficulties in reading) on female characters. Hence a textual woman can be constituted both as source of coherence and as principle of disorder, a center and a verge, damned if she does and damned if she doesn't.[21]

No comparable criticism can be found for Miranda, or for that matter for *The Tempest* as a whole, which has accumulated no comparable sacred history as a "problem play." Indeed, until recently one attractive critical orthodoxy bespoke harmony of effect rather than difficulty of interpretation: the identification of Prospero with Shakespeare, and the play-text itself with a theatrical valedictory, rendered language in *The Tempest* majestically referential, and meaning a final, paternal, gift. The play-text was used to validate, not just a fanciful (and circular) chronology for Shakespeare's production, but a critical ideology dependent on authorial control of and embeddedness in the text for meaning to be possible; as a result, the play-text's complex enactment of the rituals and suppressions of power was itself suppressed, unacknowledged, neutralized as the fictive machinations of a playwright, or a manipulator of nature. Given such a triumphant exposition of a strength all the more masterful for the way in which it represents its own (willing, and hence cancelable) abjuration, it is no wonder that such readings of *The Tempest* did not address the discursive position of those subjected to that strength—Caliban, Ariel, and Miranda foremost among them.

But what we talk about when we talk about *The Tempest* these days, is colonialism, as is abundantly familiar.[22] From fantasizing about literary and textual mastery, we have moved to proclaiming an uneasiness about the readiness of dominant modes of Western culture to undertake such mastery, and thereby to make utterance monolithic. As a result, *The Tempest* now exposes the complicity of literary discourse in the institution and maintenance of a dominant culture: it has thus become nearly an orthodoxy in itself to discuss the play as the literary palimpsest of early modern European imperialism, as a text that enacts the suppression of alternative modes of signification and erases signs of difference in a consolidation of European cultural hegemony. Even readings that consider the significance of Prospero's familial anxieties or his dynastic preoccupations reinforce the preeminence of colonialist discourse as a precondition, since they have already defined the problematics of analysis in terms of a centralizing authority seeking to expand its dominion. Thus colonialism is paternalism writ large; the benign poet-figure who puts aside his magic becomes the author of family, manipulation, and suppression.

Notable in colonialist analyses of *The Tempest* is what often goes unnoted: the gendering of the relationships posited as the site of reading. Having isolated the movements of one sort of Renaissance power—geopolitical, dynastic, imperialistic—many (not all) analysts of colonial structures have swerved away from that of gender.[23] This is not, of course, to assert that no critic interested in the discourse of Renaissance colonialism has considered Miranda (who is the only female summoned more than allusively), Sycorax or, for that matter, the typographical vestiges of Prospero's wife.[24] It is, however, to say that having excavated systems of repression and occultation lying dormant within the magic-decked space summoned by the Shakespearean play-text, these readers of the early modern colonialist subtext have inadequately dealt with the space and functions that can be gendered female. Paul Brown's essay, " 'This thing of darkness I acknowledge mine,' " in Jonathan Dollimore and Alan Sinfield's collection *Political Shakespeare,* may be taken as exemplary.[25] Miranda is figured in his essay as a partitioned object of colonialism, whose contradictory provocations of the European subject are thus split between the wild man and the timid virgin. The observation has much force, but its binarism is symptomatic in that it reduces the acknowledged feminine to anamorphic specularity, to a subspecies of a masculine itself represented as a subspecies of the European male norm. Equally, it obscures the way the text continues, and, indeed, inflects the absorption in history of female subjectivity into the hegemonic. It is perhaps less easy to see the feminine in *The Tempest* as undomesticated than it is to ignore the costs of domestication.

Witness Ariel as a special case. Arguably, Ariel is as deeply enslaved by the discourse of domination as is Caliban, his brutish—and unequivocally masculine—counterpart.[26] The spirit's efficient agency, as well as his goodwill toward Prospero, might then be taken as the response of the "good slave," a failure to oppose because of an identification with the ruler rather than an acknowledgment of what it means to be ruled. As my appositional adjectives for Caliban have implied, however, Ariel's ambiguous gender may have as much to do with the spirit's being exempted from critical scrutiny as does any willing administration of Prospero's machinations. In a sense, the two points are coextensive: when Ariel asks "Do you love me, master? No?" (4.1.48), the question signals, among other things, that emotional satisfaction is the proper response to services rendered—and it is a transaction readily gendered in modern discourse as feminine.

Indeed, Ariel has often been not only impersonated by a woman, but costumed as one. As Stephen Orgel notes in his introduction to the Oxford edition of the play, although Davenant established Ariel as a male role, from the early eighteenth century until 1930 Ariel was played as a female.[27] The illustration of Ariel-as-quasi-Victorian angel that Orgel reproduces from an 1847 Sadler's Wells production goes a long way toward explaining the lasting association between a self-abnegating utility, an impression of ethereality, and a consequent femininity that hovers around the figure of Ariel.

Still, one additional point emerges from the production history of the play, with consequences for the role that gender (under)plays in the productions of modernity with which I am here concerned. That Ariel could have been received as unequivocally male until the eighteenth century suggests that it is necessary to rethink our models of gendering. Specifically, the reception calls into question the oppositional calibration of difference so useful, and yet so compromising, in Brown's essay, and it demands, in turn, that the play be seen as a formative constituent of such oppositions.

Ariel's imprecision of gender suggests that femininity is interstitial in *The Tempest* produced by colonialist readings: it can be fleetingly glimpsed, as if out of the corner of the aesthetic eye, as the stuff that holds the play together, as the ground upon which its staging of power is wrought. Rose's critique of past readings suggests that for such readings the feminine converges on the textual, where the pathology of the former, as ascribed and represented, induces (and thus cannot but stand in for) the deficiencies of the latter. But the symptomatic analysis may go yet further to assert that the text itself, under these conditions, is deemed a species of the feminine—the matrix and ground of meaning, the mother lode of interpretation, whose innate difficulties are to be mastered by masterly readings, whose ontology is recuperable only insofar as it can be made to reproduce versions of the masculine. Hence the discursive consciousness that has informed most colonialist readings of *The Tempest,* and that has played out the drama of oppression as the drama of Caliban, nevertheless stops short of examining the means of reproduction—of textual as well as of cultural meaning. The current occlusion of gendered concerns is symptomatic not only of present reading practices; indeed, it also constitutes a way to examine the productive redeployment of gender as well as of race as they figure in representations of the civilizing process. If Renaissance humanists inscribed a domestic and civic polity through the self-sacrifice of Lucretia, Miranda embodies a newly useful association between orders of nature, and hence a way to unearth, interrogate, those symptomatic binarisms upon which Paul Brown's argument draws. Before accepting the figuration of her gender as an already legible signifier of difference, isolating her in the play-text lets us see what renders her a fitting agent for the work of discursive colonization.

Appositely, insofar as Miranda figures significantly in *The Tempest,* it is as a vehicle of the reproductive, if by that word is understood a complex of social as well as biological functions. She is the object of both Caliban's and Prospero's dynastic ambitions, a conduit for the perpetuation of patriarchal hegemony. The marriage Prospero contrives between Miranda and Ferdinand is one of the master plots of European diplomacy, the products of which union guarantee the restoration of the political order in rupture. That order, in its turn, depends critically upon keeping Miranda untouched by sexual experience, and therefore a suitable vessel for the establishment of a dynasty to rule Naples through marriage to Ferdinand.

It is a view of Miranda's usefulness given universal currency in the play. Ferdinand's first speech to Miranda inquires whether she be "maid" (1.2.428), and he later promises that Miranda "if a virgin" and hence unattached will be made queen of Naples (1.2.448–450). Caliban's attempted rape of Miranda would, of course, have rendered her unfit for the perpetuation of a Neapolitan monarchy. But as a scheme to exact revenge for his disenfranchisement, it is isomorphic with Prospero's own revenge plot. And while Ferdinand's courtly behavior to Miranda seems to juxtapose true affection to base lust and bestial violence, Prospero's anxieties about "too light winning" and the premature rupture of the "virgin-knot" (1.2.452–453; 4.1.13–23) suggest a structural analogy between the desires of the legitimate intended and the depredations of the savage. And suggests, as well, the ineradicable connection between the very matter of femininity—the space of the female body itself—and the political terrain upon which *The Tempest* is founded and which it in turn maps out. To "people else This isle with Calibans" (1.2.349–350), as the slave memorializes his thwarted ambition, is to stake preemptive claim simultaneously to two equivalent bodies: that of the island (which would become his, not by patrimony, but by a violent travesty of matrimony), and that of Miranda. Caliban, who is "all the subjects" Prospero has (1.2.341), seeks to enact his revolt against the usurper of his island upon the form of Prospero's daughter. Caliban wants the island back, and Prospero to retain control of it until he recovers his dukedom. In both cases, their contention for the island or the duchy of Milan is enacted through, and upon, Miranda. The close correspondence this betrays between the female body on the one hand, and the notion of material ground, of territoriality, on the other, is of course an enabling move both for colonization and the allied ideology of the New Science, an issue to which I return in analyses of Jan van der Straet's "America" and Joseph Hall's *Mundus Alter et Idem.*

The attack on Miranda seems to assert her importance as the apex of a triangulated hostility; but equally it may be read as symptomatic of her insignificance. In deeming Miranda a conduit for his vindictively sexual and dynastic ambitions, Caliban reduces her from agent of subjectivity to source of destruction, in an insurgency to be played against Prospero, but whose price is to be exacted from his daughter. Caliban's rage attempts to circulate through the female body to culminate in an act that strikes out at the originator of his oppression; yet in the process of finding an ultimate destination for that rage, Miranda is obscured, subsumed into the figure of her father, into the workings of authority and domination.

Miranda remains embedded within the text, indeed, constitutes its unexamined ground, precisely because she is so naturalized, so much a creature of her station that she all but disappears as an independent subject. Yet this is not wholly an issue of class determining affiliation, which is to say of her willing participation in the standards of the hegemonic elite represented by Prospero. It rather betokens a silence imposed upon the feminine both by that elite, the cultural structures out of which the text is generated, and by a critical tradition unable, it seems, to hear what is spoken. Although desire occasions Miranda to break out of her discursive imprisonment, as when she reveals her name to Ferdinand, or when she discloses her love for him, each time she immediately retreats because she fears breaking her father's commands not to speak. Her own capacity for self-censorship obviously underscores her absorption into the system of obedience and hierarchy which *The Tempest* lays out, where the female voice is, if not unuttered, then at least under partial erasure.

Who teaches language to Caliban? In Act I, Miranda rebukes him: "I pitied thee, Took pains to make thee speak" (1.2.352ff), and, since there is no textual ambiguity about the lines, the question may seem idle. Editors from Dryden onward, however, have clearly felt uncomfortable in ascribing to (implausibly childish?) Miranda so central a role in the construction of the slave's subjectivity, and the lines have frequently been reassigned to Prospero.[28] Yet the difference between Prospero's lessons and Miranda's is also the difference between reinforced dominion and the space of potential resistance. Since Miranda imposes the conditions of European subjectivity upon the "savage and deformed slave," she becomes the double agent of his domestication and his doom. When she gives him language—teaches him, as she reproaches him, "each hour / One thing or other" (1.2.353–354)—the vagueness, even randomness, of the lessons thus described can be opposed to Prospero's purposive instruction in the hierarchy of the cosmos, in naming "the bigger light and . . . the less, / That burn by day and night" (335–336). Prospero, deposed duke, teaches Caliban place and degree, uses natural categories to implicate the monster in the coercive Great Chain of Being, and by extension into the social order validated by the natural one. It is given to Miranda, on the other hand, to bespeak the monster's disruptive yearnings, and in so doing to teach him the language of sedition. By endowing his "purposes / With words that made them known" (356–357), she gives form herself, her own form, to his inchoate desires. Miranda provides matter for Caliban to work upon, becomes an ambiguous text, and thereby cannot escape becoming the object of his sexual violence. But her female corporeality also provides him with a mode of rebellion apposite for a system in which women are counters in the discourse of domination and metaphors for conquered territory.

As Miranda's role in the construction of Caliban as a speaking subject makes clear, the feminine in *The Tempest* can speak on behalf of the elite, or equally it can provide the means by which that elite can be cursed and combated, as represented by the talismanic name of Sycorax. At Prospero's behest, Miranda has brought Caliban into the world of European subjectivity, which is a precondition of his own subjection; Caliban reacts against that subjectivity and its agent both. In effect, Caliban's experience with European culture is constituted as a fall from an imaginary realm, whose perfect grace depended upon its incommunicability. Although Caliban's curse is often taken as directed at Prospero, it rather must be directed at Miranda: as she is allowed to confer the power of speech upon him, so she is made to become the occasion of loss. Perversely enabled, Caliban strikes out against the

European culture that has alienated him from his preprosperian existence; his violent insistence upon corporeality can be viewed as a way to reestablish control, to authorize himself through sexual reproduction. Yet the replicas which he wishes to beget through her by his rape efface Miranda, remove her both from the reproductive system and from the system of consciousness, as much as they expose her necessity to it. Their hypothetical children are to be wholly "Calibans," after all.

This need to authorize alterity, to insist by force on a place for that which has been marginalized by force, tropes the colonialist readings to which I have referred. Both seek to write the name of Caliban across the landscape of *The Tempest,* while relegating that terrain to unaccommodated nature; in both cases, reproductions of the masculine are the privileged sites of reading. The effacement, then, of Miranda and the feminine from many current readings of the play ascribes a ground level of unexamined materiality to the text, a materiality unaffected by recognition that the text is implicated in colonialist agendas. Confined by such analyses, femininity is still relegated to the status of a given, a state of nature. In their various ways, such readings localize the colonialist agenda of the play-text to a singular historical event: England's first encounters with the New World. Implicitly, they claim that, insofar as the text is a production arising out of a particular moment in culture—the emergence of British interest in North America—what needs to be recaptured is its staging of the ethnic Other, its suppression of racial and cultural difference. But of course there is more than one subjugated discourse, more than one way to rip through the apparent integrity of Prospero's island kingdom. For that matter, there are ways other than the interrogation of subjectivity to open up *The Tempest* to the subtle interlacings of historical specificity, as a consideration of the making of woman's colonial body makes clear.

How then to redraw the map of the colonized so that it encompasses not only categories of race but of gender, not only Caliban but also his counterpart and foil Ariel, so that it encompasses Miranda as well? The answer lies in a reshaping of the agenda pursued by colonialist readings, by a displacement of the (masculine) subject from a position of centrality and a consequent emphasis on the way in which culture becomes textual matter, where in fact the very question of materiality and its relationship to textual practice gains new force from historical circumstance. In one sense, the colonialist readings that have proliferated, and that have focused on Caliban, can be read as attempts to dismantle the formation of "the natural" and all its attendant hierarchies, to break down the ostensible given into the social category whose signs of production have been effaced. Thus Caliban's problematic commingling of savageness and sensibility betrays at once the operation of an emergent racial taxonomy and the undermining of such taxonomy. Then it becomes apparent that what has operated in such recuperative readings is a reinscribed set of hierarchical categories, which gives discursive place to one form of nature recalcitrant—Caliban, the deformed and savage slave—and leaves comparatively unexamined nature domesticated, pliant, and serviceable—Ariel the "tricksy" spirit, and Miranda, the cynosure of all eyes.

It is in fact this very pliancy that makes it appropriate to read her, too, as an embodiment of the natural, even given her superior access to language; after all, relegating something to the status of background, even in the literal sense of the term, is an operation fraught with the tacit adjudications of ideology. In that case, her malleability and lack of autonomy are productions of an ideal: not solely of womanhood, although certainly that, but also of the view of nature which underwrites all colonialist ventures and which makes them coincident with the emergence of scientific modernity. If *The Tempest* bears oblique witness to the durability of an equation between woman and nature, that equation is not exhaustive: "nature" as discursive production pervades the text, encompassing the island, its storms and its noises, the way its forces are marshaled and embodied. Against its instantiations— whether minatory or domestic, whether raw like Caliban's berries, or as suavely concocted as the feast Ariel so abruptly snatches away—Prospero and the other white male Europeans ground themselves, and themselves run aground.

Miranda belongs to that island; indeed, given her extreme youth at the time of her father's exile, she belongs to it at least as much as she does to the world of the court, her skill at chess notwithstanding. Even more, the representational practices of the European Renaissance position her as that island, and the social practices ensured that she did not remain uncolonized.

Admiring Miranda thus becomes a two-edged activity. On the one hand, it misleadingly suggests a validation of the play's positioning of the female: the enforced object of enforced admiration, as her very name attests; a mirror, a cynosure, the gazed-upon by compulsion. But on the other, it casts attention on a system of approbation that pervades *The Tempest,* where behest and obedience are necessary to the smooth unfolding of Prospero's master plot. Within this system, the "admiring" of my title becomes an adjective, one that moreover carries over from Miranda to Ariel, the "tricksy spirit" without whose administrations Prospero is unable to enact his control of nature.

The reciprocality, or rather polyvalence, of the verbal formula invoked to caption this argument suggests that the activity is as important as its nominal marker, that admiring is as much to the point as what is admired. Yet looking, which is the physical basis of admiration, is also the cultural trope by means of which Western culture, from the Renaissance to the present, has figured its interest in nature, and its intention to harness that nature into socially profitable forms.

The Tempest is a tricky play. Given the aesthetic satisfaction often generated by its conclusion, the ultimate contriteness and docility of the agents of the natural in the play effect a mystification of the intransigence of nature at this early moment in the formation of the scientific. It can be seen just how neat the play's closure is, its resolution of conflict within the experience and representation of the New World, by extending the field of analysis.

One evidential basis for exploring the connections between the feminine, the American land, and the colonizing of nature is provided by Jan van der Straet's 1600 portfolio of engravings entitled *Nova Reperta,* or "New Discoveries," an anthology of contemporary inventions, discoveries, and technical innovations.[29] The discussion in chapter 1 of Harriot's *Briefe and True Report of the New Found Land of Virginia* took the body of the Algonkian, brought into relation with that of an ancient Pict, to argue for the emergence of two related protomodern discourses: an implicit historiography that must bring difference into line with a universal narrative of apprehensibility, and an implicit ethnography that, while sharing the universalist agenda of history, deems some bodies to signify more than others. Thus, the "sciences of man," even in a preformational state, are readily understood as a species of discursive colonization. When it comes to the similarly colonial science of nature and its work of objectification, a universal of another sort is ready to hand for reinscription—the body of woman.

Van der Straet's collection is introduced by a title page on which the cartouched "Nova Reperta" appears to be balanced atop a printing press; a line of printed pages hung to dry also depends from the cartouche (figure 6). To the left, a nude woman, with a cape slung over her shoulders and a cuff on her upper right arm, steps forward. She holds a snake that has bent itself into a hoop by biting its own tail, while she points to a circle enclosing a map of the Americas, around which a legend reads: "I. Christophor Columbus Genuens. Inventor. Americus Vespuccius Florent. Retector et denominator." To the right is what appears to be the diagram of a compass (or magnetic wind rose) whose north point is occupied by the Roman number II, and whose south point is designated by a triangle of three stars. Around it reads the legend: "II. Flavius Amalfitanus Italus Inventor." To the right of *that* circle, an old bearded man wearing a long shift and carrying, like the woman, a snake hoop and a pointer (although in opposite hands to hers), seems about to leave the plane of representation.[30] Beneath these complementary pairs is an array of inventions and objects: a still; some logs labeled *hyacum,* or guaiacum, a New World tree that when distilled provided a tonic to relieve the symptoms of syphilis; a cannon (centrally located) that betokens the invention of gunpowder; a mechanical clock; a saddle displaying stirrups; a mulberry tree with silkworms. (All these inventions, and others, such as the astrolabe and the magnet, are presented

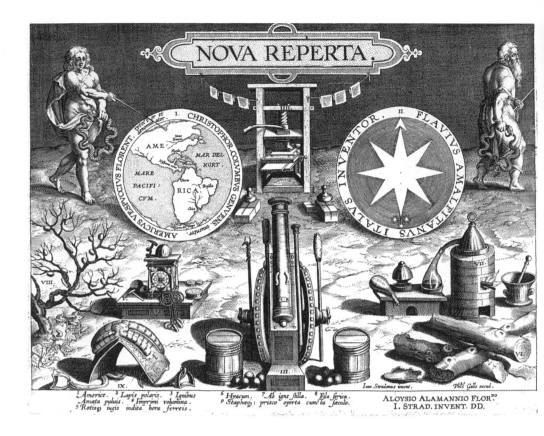

6. Jan van der Straet, frontispiece to *Nova Reperta* (by permission of the Folger Shakespeare Library).

individually in the plates that follow.) It is tempting to think of the young woman and the old man in terms of a general cultural schematic, an emergent opposition between new (world) and old, between novelties of manufacture and discovery and the knowledges that they seem to supplant, which are nowhere else represented on the page.

But it is the first engraving after the title page that particularly interests me, an image that floats from edition to edition, and so capable of being redefined according to context. While Michel de Certeau places the image in a 1619 edition of *America decima pars,* Louis Montrose seems to refer to "America" as a single plate produced circa

1580.[31] Thus their readings take the single image as emblematic of a particular set of colonial relations embodied in, and made apprehensible through, gender. In the particular collection I am considering, the gendered encounter depicted in "America" also proposes a link between discovery and invention, following the agenda of the title page. It is significant that van der Straet commences *Nova Reperta* with such a representation of the encounter between Europe and America, and so he images modernity as a gendered meeting of male observer and female—landscape? inhabitant? embodiment of both, perhaps, combined? If on the title page the comparatively aged male figure—who may as Father Time signify the culture of antiquity—seems to be driven off by the assertive stance of the female native, other represented aspects of the encounter redress the imbalance, attempt to put femininity, novelty, nature back in their proper places. The inscription encircling the title page map implicitly suggests as much: America is contained by the names and birthplaces of the European voyagers who discovered and named her. Manifestly, there is contiguity, domestication, in the relationship between old and new.

It remains to be seen, however, whether that domestication is thoroughly accomplished by the engraving that van der Straet dedicated to the symbolic encounter (figure 7). Certainly the appearance of Amerigo Vespucci at the left side of the plate, just disembarked from the ship visible at the image's margins, suggests the indomitability of the Old World: armed and armored, erect, bearing the standard of Christianity and the tools of navigation. Yet Vespucci cannot be mistaken for the counterpart of the old man about to vanish from the title page. Instead, the explorer's encased, occluded body can be read as an emblem of the prosthetic agent of scientific and technological modernity, the colonizer of nature. Thus exempted from the symbolic obsolescence of the old man, he confronts his opposite, a naked woman who has been reclining in a hammock.[32] The shock of the encounter—which is to say the precision of the inverted relationship between them—has caused her to rise up slightly and to regard him with what seems to be startled amazement; according to the subscribed legend, she has been sum-

AMERICA.

Americen Americus retexit , &   *Semel vocauit inde semper excitam* .

7. Jan van der Straet, "America" (by permission of the Folger Shakespeare Library).

moned into a state of permanent wakefulness. She is America, his newfound-land, embodied in a woman whose undress approximates a state of nature as it confirms the exploitability of both vehicle and tenor.

While the trope of woman as landscape pervades colonialist documents, the surrounding pictorial space sets this particular meeting in a system whose complex bifurcations extend to the historicity of *The Tempest* and its versions of the natural.[33] For instance, scattered throughout the engraving's plane are representations of indigenous flora and fauna, whose very presence argues the connection between encounter and observation, between the discovery of the new and the subjection of it to intense scrutiny. The ideological position occupied by observation within emergent natural philosophy is implicit in Vespucci's keen probing of the woman before his eyes: hence the strategic location of this engraving at the beginning of a collection on proto-scientific developments. The scene designates a foundational moment in the relationship between men and nature, in which "the natural" becomes a taxonomic category that enables both the growth of technical knowledge and the concomitant discursive estrangement of the male scientific observer from the phenomenal world. "America" both comments upon, and differs from, many of the images to which it stands as a prelude, for it does not focus on human—masculine—activity within a domesticated enclosure, a machine-filled space evocative of nascent modes of production. Rather, it is the female figure who is complexly positioned in this site; she is both the rare bird of the exotic landscape, yet another example of indigenous fauna, and the point of contact for the European male explorer, thus imaged as his double, his reflection.

Despite the apparent prosthetic mastery of Vespucci, however, his conquest is not entirely guaranteed. To understand the combined recognition and menace that characterize the specular dyad of Amerigo and America in van der Straet's engraving (a dyad almost too neatly underscored by their congruent names) necessitates the intercalation of desire for this dual object within the structures of history. The discourse is now well-known: longing for the paradisiacal luxury of the New World competed with a sense of the dangers that world concealed, and the mastering of nature could be couched in the rhetoric of servitude—a contradiction written across the form of a potentially menacing woman. That the hammocked native seems quasi-Amazonian is betokened not only by her motility or by the club resting near her, but by the disturbingly gendered cannibalism being enacted in the background. It appears as though America's (mostly) female companions are in the process of devouring the strategically impaled haunches of some enemy—perhaps a hint of some prior encounter. Montrose suggests connecting up the imaged feast with an incident described by Vespucci in which a young male Spaniard who has aroused the interest of some native women passes quickly from being an object of visual, to an object of culinary, consumption.[34] The encounter, however, seems to me less urgently a reference to "an oral fantasy of female insatiability and male dismemberment" (5) made concrete than a more general, and less anxiously Freudian, staging of the contrary provocations of the natural. This is, after all, a universal Woman, and so a Nature soon itself to be made universally apprehensible through a cautious and informed study of her ways and appurtenances. If, as Bacon dictated, nature must be obeyed to be commanded, the contradictory attitude to be teased out of this articulation of his philosophical agenda finds a visual equivalent here in the gendered terms of vulnerable flesh. The apparently voluptuous land, the seemingly pliant natural world, becomes not only the mirror of European desires, but the potential for their inversion; the possibility of such betrayal is implicit in the tense opposition, and implicit as well in its conjurations of gender. By comparison, the successive engravings of novelties seem like propaganda, or compensatory attempts at mastery.

Joseph Hall's dystopian travelogue of 1605, *Mundus Alter et Idem*, further works the Renaissance connivance between the unknown's heart of darkness and the potentially monstrous regiment of women.[35] The text generally represents the New World as a series of excesses beyond the constraints of European culture, all the while rendering those excesses as travesties of European desiderata: regions whose national order is extreme drunkenness or overeating, for instance, suggest a Cockayne gone more than usually awry, an alien state where even the governing force of bodily limits, of satiety, does not exist. Nestled in among these pleasures made other, made terrifying, is New Gynia, also known as Viraginia. Clearly, the name is a parody of the cultural talisman that is Elizabeth's name, but also certainly of the ideological work the virginal designation performs within the discourse of colonization. Rather than a site of purity and infinite exploitability, the female territory of New Gynia is profoundly destabilizing to the knowledge-seeking agenda of the male subject.

By the logic of grotesque excess that characterizes the text, this gynocracy, like other presumed New World cultures encountered by the narrator, provides a narrative of escalating lawlessness: it moves from a stereotypical, unbounded femininity (woman as lugubrious, or garrulous, or lubricious) to woman as, perhaps, no longer woman at all. Accordingly, the itinerary of Hall's text passes through a leaky landscape of tongues and tears (Linguadocia and Ploravia, in the original Latin), through Erotium, a region of sexual abandon; from the Hermaphroditic Isle to Amazonia itself, a garrison city. In so doing, it charts the progress of anxiety before the letter and structure both, as the feminine moves from mild incontinence to sexual license, from gender ambiguity to the ultimate transgression of assumed masculinity. Hall's text had attempted to contain the disruptive power of the alien natural (and largely male) body by burlesquing it through traditional representations of excess and bestiality. When it comes to gendering nature through writing of women, however, the text is laconic, suppressive, unable to speak at length, unable to position its avatars of femininity along the spectrum of inferiority it works out for the brutish and appetitive.

Van der Straet's engraving equates the hostile yet seductive New World with the wary Amazonian warrior; Hall's text, on the contrary, implies that the equation is potentially commutative, that the female body is an exemplary site for innovation. Indeed, the partition and scrutiny of the socially constituted female body undertaken in *Mundus Alter et Idem* finds a peculiar resonance in the practices of early modern dissection exemplified in Vesalius (figure 8).[36] While *De Humani Corporis Fabrica* is a foundational text for a discourse of biological truth made visible, its illustrations incompletely rationalize the bodily structures that they expose to view. This incompleteness is never more visible than with the text's few renditions of women's anatomy, where representations of female genitalia as penile infamously reveal the persistence of Galenic dogma in a testament to empirical knowledge of the body.[37] More broadly, the illustrations reveal that coding even of the body as a fact of nature is nevertheless haunted, as Hall's text is, by the social constructions that precede it.

Yet perhaps the frontispiece—which depicts an anatomy theater

8. Title-page to the first edition of the *De Humani Corporis Fabrica* of Vesalius (from the Rosenwald Collection, Library of Congress).

whose various denizens cluster around the corpse of a woman being dissected by Vesalius—is more to the point here. Actual dissections of female corpses were rare in the sixteenth century; judging by the symbolic paraphernalia to be found at the site, however, the scene depicted in the frontispiece is offered less as a transcription than as a moment of ideological clarity. It is as if the male body, no matter how often it served for the concrete practice of anatomy, were not so readily a sign of a nature to be domesticated through emergent structures of power-knowledge as the passive and inert body of a woman, with tongue silenced and orifices but the conduits of knowledge. Hall's instantiation of the alien colonial space as the opened-up bodies of women offers in words a version of what the frontispiece offers pictorially—but with difference not as threat but as instrument. When taken, then, with van der Straet's engraving as well as the one that accompanies Vesalius, Hall's text registers woman as double counter, both for the object of colonialism and for nature as object. In discursive practice, the New World and the world of phenomena can only be assimilated to the nascent scopic regime through sexing, through insertion into the semiotic field of normative gender relations. Women's bodies thus provide the site upon which certain dominant meanings are built and rebuilt: they are pieces of nature not wholly relegated to the background, to the operative nexus of the material.

But the further point is not to essentialize the relationship between woman and exploration (whether colonial or anatomical), or to show the vectors of symbolization going all in one way. This would be to reproduce meaning from a feminine body taken as "natural" in itself, and so to reinscribe the very process whose working can be detected in *The Tempest.* As I suggested, Hall's text argues for a commutative production of significance, with the colonial experience seen as dystopian because it codes that experience through woman, and yet it is also antifeminist in its overt, and culturally proximate, positioning of woman as other, as natural. When the *Mundus Alter et Idem* presents woman as American landscape, it is to master the terrain, to stop up the orifices, to keep the female in bounds—a counterturn that implies a prior turn, that implies the mobilization of misogyny in dialectical response to the destabilization of precedent social hierarchies. What those destabilized structures might be cannot be answered through empirical modes of presentation, themselves complicitous, when filtered through the scientific, in the emergence of this differential semiosis. But they can be glimpsed through their effects, through the textual fallout of the sort with which I am concerned. The colonial encounters I have discussed participate in the construction, and hence the making available, of "woman" as a historically activated category of analysis that can then be circulated through the cultural networks of early modern Europe.[38] For instance, the connection of women to water, the stereotypes of feminine talkativeness, sexual license, even of aggression (to speak once more of the Amazon) are quickened by their insertion into the discourse of Renaissance colonialism, which had spoken of the "maidenhead" of an unsacked Guiana and had named a lushly overgrown and powerful river in America after a mythical tribe of warlike women whose home, finally, had been located.[39] Through such particularities, woman becomes the matter of the late Renaissance.

The move from America back to Miranda exchanges the unaccommodated forest for the cultivated garden, substitutes for wild difference an abashed control; in a motion at least partly antithetical, however, it also exchanges specularity for compelled admiration, if naming constitutes symbolic necessity. When the figure of America is constituted as the repetition and doubling of the masculine order, the identity is foundational for the drawing in of raw matter, which is to say raw difference, raw experience, to the confines of preexistent form. The initial strangeness, even discursive intransigence, of the colonial object is a function of its representational ambiguity: thus the sexy woman is ominously masculinized. When, however, the woman in the colonial space is always in danger, as Miranda is, of being taken as the background/backdrop for the primary drama, it is as good as a promise that drama—whether textual or historical—will be resolved in favor of the hegemonic. So potent and yet so effaced is the symbolic force of her domestication that gender all but disappears as an overt category of analysis in *The New Atlantis.* There, it is More's *Utopia* that gets reconfigured into an engine of modernity: perhaps ironically, given all of humanism's productive mobilizations and suppressions of femininity, it is the humanist text that does the woman's work.

3

The New Atlantis and

the Uses of Utopia

The writings of Francis Bacon occupy a liminal position in the canon of literature written in the English Renaissance. It is possible to talk about his synthetic projects piecemeal and to draw lines of inclusion and exclusion: the *Essays* are included, as are *The New Atlantis* and portions of *The Advancement of Learning*. As the texts draw more deeply into the project of constructing a new horizon of knowledge through the study and manipulation of nature, however, they have proven less useful to the construction of Literature as a discursive edifice: hence most of *The Great Instauration,* meant by its author to stand as the foundation of knowledge reconceptualized, is extramural.

That the literary canon does not contain much scientific matter, and specifically that some of Bacon's works fit and others do not, may seem parodically obvious. (Why the canon has been so constituted, however, is another matter.) But from the vantage of nearly four hundred years it is in part Bacon's writings that allow us to make it. Not that the texts in themselves map out a discursive essence of science or of literature, available for the taxonomy of transcendence: rather, in varying degrees they inaugurate the process of cultural differentiation that has marked off literature as the aesthetic and the scientific as the productive.[1] To say as much is perhaps to be guilty of another banality: seeing texts as both palimpsests and predictions that encode the conditions of their emergence and that alter those conditions as a function of their historicity. Yet the history encoded by Bacon's writings is highly pertinent to the politics of canon formation, insofar as that politics in its turn depends upon a prior ideological move, one that effectively removes literature from a network of cultural signification to position it, instead, as an autonomous system of transcendent aesthetic experience, an alternative to quotidian demands on time and industry.[2]

While I am aware of collapsing a complex historical argument whose most concrete instantiations occur in the nineteenth century into a few compact assertions that stretch back into the early modern world, such assertions are a necessary, if provisional, preamble. Bacon's texts are crucial for understanding discursive categorization because they mark the movement from late Renaissance humanism, with its investment in *litterae,* in literary texts as the ground of culture, to the coming-into-publicity of the early modern world, with the concomitant positioning of nature as primary text, and with modes of quantification and classification as increasingly dominant practices. Allowing for the nonlinear transformations of culture that interpose themselves between the present site of judgment and what I have posited as the predictiveness of the Baconian text, it seems fair to say that the present highly marked differentiation between literary forms and scientific forms is a distant consequence of Baconian reworkings of humanism.[3]

The argument needs to be cast less deterministically, and in more historically nuanced terms. Much work has been done in English Renaissance studies concerning the complex filiations of power and patronage between humanists such as More, Sidney, Spenser, or Jonson, and their monarchs, especially in arguing that their writings negotiate the possibility of critique from within poetic conventions ostensibly governed by the necessity of praise.[4] But the contents of such arguments, powerful though they are, are less significant in this context than the fact that they necessarily ground themselves against an understanding of humanist literature as political discourse. In various ways, they identify an allegiance between the court as a center of authority and humanism as its mode of representation and legitimation. The Baconian text produces difference from within this model by inserting itself into the flow of power between monarch and representation. It reassigns humanist modes to its own agenda, the propagation of knowledge-from-nature, and hence it directs courtly interest away from a poetry reconfigured as false, iconic, and unproductive to a truth derived from commanding the natural world.

In 1608 Bacon prepared a brief for King James to encourage the "plantation" of Ireland.[5] In its course he draws a comparison between its proposed structure of governance and that of another enterprise to hand, one but fitfully pursued: "The second [proposition] is that your Majesty would make a correspondency between the commission there, and a council of plantation here. Wherein I warrant myself by the precedent of the like council of plantation for Virginia; an enterprise in my opinion differing as much from this, as Amadis de Gaul differs from Caesar's Commentaries" (4.123). The advice here is pragmatic and direct—but takes a strange detour by likening Jacobean imperialism to written, indeed, literary texts. As the reference to Caesar suggests, Ireland is "another Britain" (119), and the British may take Caesar's part in a war represented as strategy, its impetus and hence its ontology linear.

Amadis de Gaule in this context seems to define itself by opposition. A massive romance written in Spain by at least five different authors, then amplified in France before appearing in England, it suggests a range of alternative values: not the tight focus of Roman imperium but the ambition of early Renaissance colonialism, not linearity but narrative brachiolation, not chronicle but romance. And thus Virginia, too, is defined by opposition: it is a complicated text already inscribed by the hand of European incursion, but it is not yet entirely the site of conquest. Instead, it becomes, in Bacon's analogy, a stage for imaginative transformation, for fantasy, for fiction.

As Bacon's comparisons indicate, social and political enterprises can be recast by means of fictions. To impose fictional form on the world of actual practices is to render those practices at once formal and ideal, and thus to cloak them in a coherence the more useful for literature's complex and oblique relationship to the production of ideology. Although the domain claimed for poetry by Renaissance theorists is the aesthetic and the didactic, in such cases as this one, its form can be invoked preemptively, even authoritatively, to legitimate hegemonic practices. This is precisely the agenda of Bacon's discourse. Written to persuade the king that Ireland can be subjugated to English needs, it produces the literary both as a warrant for settlement and as a blind for the real acts of violence such settlement necessitates. In the lines I have initially quoted, Virginia, too, figures as a fiction, as the romantic narrative of colonialist ease, to whose plot one more episode, one more encounter, can always be added. The open-ended form of romance at once authorizes and mystifies the work of material domination that Bacon proposes, and that open-endedness provides aesthetic distance from the world of violent colonialist practices. This disavowal of the need for closure, this preemptive use of fictional form, permeates *The New Atlantis*. It enables the scientific to be produced under the ostensible sign of the literary, and it results in an act of discursive colonization that tropes the cultural agendas of Jacobean imperialism to which I have initially alluded.

Of all Bacon's texts, *The New Atlantis* can most readily be categorized as fictional. Despite its romantic aspects, however, it seeks alignment with More's *Utopia* rather than with *Amadis de Gaule*, which is to say with a political fiction that is also politic in its fictionality. *Utopia* maintains a difficult balance between revolutionary ideality and practical impossibility, simultaneously affirming and denying its radical agenda. More's text operates by invoking specific historical conditions in order to resolve the cultural anxieties attendant upon those conditions; at the same time, it engages ideological premises without overtly displaying its own ideological legitimation. Hence the paradox of More's *serio ludere*. In discussing Louis Marin's elaboration in *Utopics* upon the bipolarity of More, Fredric Jameson has noted the tendency of the text hallmarked as "utopian" to concede—indeed, to thematize—the conditions of its own emergence: "All genuine Utopias betray a complicated apparatus which is designed to 'neutralize' the topical allusion at the same time that it produces and reinforces it."[6]

Thus defined, the utopian fiction problematizes the culture out of which it emerges, throws the forms of power it ascribes to that problematized culture off balance. What has been called More's communism, for an obvious instance, undermines the ground of the Henrician state and thereby makes obsolete prior systems of material worth or social privilege. But it is a revolution on paper, contained within (and by) the purview of the text. To put it less starkly, the manifold nodes of impossibility and opposition that Marin details within More's text guarantee that an overt formalism prevails over its engagement with ideology. This formalism (which may conceivably be inscribed within Renaissance humanism proper) eschews hortatory social agendas in offsetting critique with textual play by loosening the bounds of representation.

Yet its potential capacity to allow for the rehearsal of specific historical situations means that utopian form can be directed away from the production of paradox. In that case, the equipoise maintained by this formalism between the acknowledgment (relief?) of circumstantial anxiety, and the recognition that the social remedies it proposes for that anxiety are impossible, beyond its ken, can mutate into a simulacrum of utopian form. It offers the ideal it proposes as obtainable, and ultimately it transforms texts into social machines in a way radically different from Renaissance notions of *dulce et utile.* A text "mutated" in this way would bear the apparent (formal) signs of the utopia, but that text would abandon the discursive stasis that the *Utopia* possesses and presents as a condition of its own existence. If the ideal embodied by a utopian text can be achieved as social practice, not just apprehended as a formalized solution, the mechanism for change from the present imperfect to the future perfect must partly inhere in the narrative. Certainly, most of Bacon's writings concerned with propagandizing the scientific offer themselves as operants of the type I have sketched: their efforts to repair the defects of the Fall may well bespeak profound (and maybe prerevolutionary) nostalgia, but within the Baconian program they are also a call to arms—or at least to research.[7]

This, too, is the case with *The New Atlantis.* Insofar as it promotes Baconian philosophy, and thus works to produce in actual practice what it summons in the representational, the text reconfigures its own medium, invokes the literary only to present itself as utopian (a better word might be progressive) social agenda rather than as *Utopia*n text. In introducing such philosophical inquiry as the ground upon which society is to be erected, the narrative—which depends upon the cultural role of literature, the legacy of humanism and, indeed, of More, for its force—seeks to replace the self-conscious fiction with the empirical structures of emergent science. *The New Atlantis* belies the *Utopia,* even as it invites juxtaposition with it: unlike More's, Bacon's text is ready, as the Virginian colony is ready, to take on all voyagers, to convert all readers.

I have suggested that colonialism is more than a thematized element within *The New Atlantis:* it is a defining move in the emergence of modern scientific practice from within late Renaissance culture. The process of resorption and transformation performed by the text with regard to its humanist authority becomes a discursive reproduction of European practice in the New World. The connection between the so-called New Science and the New World is repeatedly urged in the sixteenth and seventeenth centuries, not only within the Baconian program, but within Continental natural philosophy. Expeditions to America and scientific programs both propagandize themselves as voyages out into uncharted territory, where the sense of excitement that attaches to new ventures covers over the work of domination that underwrites exploration of the globe and of nature.[8] The importance accorded to novelty, and the closeness of the scientific enterprise to the colonial one, is attested to by the frontispiece to *The Great Instauration.* In the foreground of the engraving a lone ship sails out through the Pillars of Hercules, the boundaries of the known world and hence symbolic of known experience. The eye is thus invited to focus on a moment of transition, poised between the old and the new and informed by both; this ship so engrosses the attention that another ship at the far left, poised on the threshold of representation, can barely be discerned. This other ship's position on the page marks, in turn, the position within late Renaissance culture, and its dominant forms of meaning, of the Baconian enterprise. The engraving reproduces a moment of cul-

tural liminality: a movement, discursively posited rather than phys-
ically apprehended, at the horizon of experience. Beyond that horizon,
experience itself becomes novel—becomes, indeed, experiment.[9] The
distant ship is thus the palimpsest of prior texts as they make themselves felt in *The New Atlantis.*

The analogies between frontispiece and text, colonialism and natural philosophy, insist, at base, upon culture as production and as representation, and upon the need to see unprecedented experiences and to
devise unprecedented structures of knowledge in terms of prior apprehensible form. Further, these analogies insist on the "literariness" of
that form. As I have suggested, a term such as "literariness" is to be
used with cautious prolepsis, since literature as profession and as cultural production is itself in simultaneous emergence; nevertheless, a
sense of literary narrative pervades travel accounts, as it pervades Bacon's discussions of colonialism.[10] In his texts, and most especially in
The New Atlantis, colonialism becomes both topos and trope, a culturally available validation of the novelty proclaimed by his philosophical
program. But the relationship is not linear, since *The New Atlantis*
recasts the work of colonization parodically, through an inversion governed by its utopianism. It recounts a voyage to an unknown land
where, as with More's ideal citizens, the natives are presented as superior to the newcomers in religion, governance, and knowledge. Bensalemite knowledge, based on a fetishized factuality, excludes the artificial mendacity of fiction from its ken. Yet Bacon's text promotes this
system through utopian narrative, alienates that narrative in effect
from its originary positioning. With *The New Atlantis,* the discursive
terrain of humanism, dominated as it is by *litterae,* by literature, is
appropriated by the emergent ideology of science.

Through the warrant of its colonized utopianism, *The New Atlantis*
also reproduces the culture of Jacobean England, transformed to fit
the ideological contours of the scientific. Just as the encounter with
previously unknown civilizations in the Americas necessitated rethinking the stipulated plenitude of the Old World, so does the imagined
existence of the Bensalemites provoke a reexamination of English

seventeenth-century society.[11] The space both created and cleared by the premise of a utopianism based on natural philosophy changes the ground of authority: religious, civil, and, ultimately, monarchical power are occluded and displaced by the inquisitional, by the power to probe nature. The redrawn lines of authority derive from the specific character of the Baconian enterprise, in which the capacity to see accurately is the mark of social and moral eminence. This ability to survey nature constitutes the dynamism that unbalances the utopian, and that determines both Baconian praxis and the narrative agenda of *The New Atlantis*. In this text, seeing is more than believing: it is knowing, and that knowledge, coextensive with the act of reading and generated by it, brings the Baconian ideal into the arena of potentiality. If, as I have initially suggested, fictions provide the means by which cultures engage in a complex and ideologically laden form of re-formation, then *The New Atlantis* attests proleptically, if not telelogically, to the moment when cultural fictions give precedence to science.[12]

I have indicated that the romantic open-endedness which sees Virginia as *Amadis de Gaule* has a bearing upon the realm of material practices. But it is not only written forms that indicate Bacon's stake in the structures of desire in which the New World was enmeshed. In 1609 he became a shareholder in the Virginia Company and a member of its London-based council. If at times the New World was to bear the freight of romance, Bacon was one of those who saw the fiction's more prosaic economic potential.

A brief rehearsal of *The New Atlantis* makes clear the pertinence of a nascent imperialism constellated with an equally nascent empiricism. Spanish sailors attempt to sail for China and Japan, recapitulating the voyage—and in some sense the fortuitous miscarriage—of Columbus. When they lose their course, as he did, they find Bensalem, the "New Atlantis," instead. Once there, they are made familiar with the habits of the Bensalemites, most especially its College of the Six Days' Works (also known as Salomon's House); this institution, designed to promote the acquisition of philosophical knowledge, is the textual realization of the academy that Bacon wished James I to patronize.[13] After describing the college, the narrative breaks off, "unperfected."[14]

The nationality of the sailors, and the miscarried itinerary they pursue, inevitably suggest that part of *The New Atlantis*'s philosophical agenda is accomplished by recasting the discovery of the New World. Yet no sooner is the identification of New Atlantis with America made than it begins to rest uneasily. The New World is not truly "new" at all, being but "the great Atlantis (that you call America)" (3.141) rediscovered. Further, its very novelty as discovery is called into question, since Atlantis/America was known to all the ancients and continued to have been known by the Bensalemites, their closest neighbors and the purported contemporaries of Bacon's English audience. Indeed, the inhabitants of America have not always been so uncouth as European explorers had found them, as the governor in charge of the newcomers tells them:

the great Atlantis was utterly lost and destroyed: not by a great earthquake . . . but by a particular deluge or inundation. . . . But it is true that the same inundation was not deep; not past forty foot, in most places from the ground: so that although it destroyed man and beast generally, yet some few wild inhabitants of the world escaped. . . . So, as marvel you not at the thin population of America, nor at the rudeness and ignorance of the people; for you must account your inhabitants of America as a young people; younger a thousand years, at the least, than the rest of the world. . . . For the poor remnant of human seed which remained in their mountains peopled the country again slowly, by little and little; and being simple and savage people, (not like Noah and his sons, which was the chief family of the earth,) they were not able to leave letters, arts, and civility to their posterity. (3.142–143)

This passage is in some ways reminiscent of the juxtaposition between Pict and Algonkian in Harriot's *Briefe and True Report of the New Found Land of Virginia:* both insist on the juniority, indeed, the infancy, of indigenous American culture, and both connect that infancy to a lack of permanent signification. Yet Bacon's text modulates the chronological trope of universal development. While its inferior position-

ing of American natives affirms an enabling, Eurocentric racial hierarchy, the very breadth of information provided in the passage shifts
discursive emphasis from the European to the New Atlantan. The Bensalemite governor's disquisition is both authoritative in tone and comprehensive in scope, providing a corrective to ancient mythology, inte-
grating the New World into the biblical world by means of allusion,
and locating the development of its culture in the schema of universal
history.[15] What renders this Bensalemite's narrative of origins so significant within the text's larger framework is the evident superiority of
his own society—not just compared to that of America, but to that represented by the Spaniards. The easy assurance of the governor's cultural
determinism reverses the logic of colonial discourse practiced by the
Spaniards and the English alike: as the denizen of a new America, that
governor should be the object of such possessive scrutiny rather than its
commanding subject. Here, the closeness of America to Bensalem—by
the force of the geography in the text as well as of the historical models
that inform it—is the sign of opposition. The reversal is key to understanding the inflection given to colonialist domination within the text:
the oppression of otherness which is elsewhere the focus of colonization is here transformed into a visual interrogation and transferred to
a native population more suave than the Europeans who "discover"
them.

From the moment of the first encounter, it is, instead, the Spaniards who are scrutinized, contained, and regulated by their apparently
benign hosts. When they first try to come ashore, the sailors are warded
away: "straightways we saw divers of the people, with bastons in their
hands, as it were forbidding us to land" (3.130). When they are finally
brought ashore they are kept at close quarters by an officer: "lifting up
his cane a little, (as they do when they give any charge or command,)
[he] said to us 'Ye are to know that the custom of the land requireth
[that] you are to keep within door for three days'" (3.133); and even
when this ban is lifted, they can never go more than a mile and a half
from the city (3.135). They are also told of "laws of secrecy concerning
strangers" (3.136, 3.140) that will render any discursive exchange but

fragmentary, and are always controlled by the Bensalemites, who because of secret envoys to Europe know more of Europe than the Spaniards can ever know of them. The seamen are aware of the constraints
under which they have been put and attempt to discern the reason
behind them. They know they are the objects of a detached scrutiny, a
desire to penetrate appearances, that has turned into a form of surveillance: "For they have by commandment (though in form of courtesy)
cloistered us within these walls for three days: who knoweth whether it
be not to take some taste of our manners and conditions? and if they
find them bad, to banish us straightways; if good, to give us further
time. For these men that they have given us for attendance may withal
have an eye on us" (3.134). The Spaniards, surveyors of the New World,
have become the surveyed, objects of the ethnographic—and colonial—
gaze.

What do the Bensalemites watch for? And by what warrant? To answer
those questions requires understanding the ethnographic structures
of knowledge and authority summoned by the text's version of the
colonial and utopian agenda—to comprehend, in short, the rhetorical
distance traveled between Bacon's Jacobean England and its fictional
counterpart. The fiction that a voyage of discovery is being reproduced
is aligned both in form and practice with Bacon's scientific program, a
point to which I shall return. More immediately, the fiction takes the
constituent elements of the Jacobean social realm and intercalates them
within the newfound one by mapping out analogous institutions in
Bensalem. Forms and observances are subtly transmuted, distilled
through the operation of a fictional matrix into components informed
by the social and political practices of the Jacobean state and yet ideologically congruent with the investigation of nature propagandized by
the text. The received coexists with the projective, the old and European with the new and scientific: the resultant version of Bensalemite
society is at once familiar and radically estranged.

Consider the Bensalemites' Christianity as an exemplary case of
estranged familiarity. It first serves in the text as yet another instance of

reversed colonialism, with the natives' conversion an already accomplished fact—and at that by revelation, the very antithesis of the laborious and linguistically incoherent mission of salvation usual in the period. Even before the papal edict of 1622, which mandated the *Congregatio de Propaganda Fidei*—and which thus expanded Catholic dominion into the New World as a counterbalance for the burgeoning Protestantism of the Old—the Spaniards had invoked the church as warrant for their colonization of America. And as the title of Samuel Purchas's 1625 collection *Purchas His Pilgrimes* suggests, the English, too, connected the annexation of territory with the conversion of souls: to refer to explorers as "pilgrims," and to their journeys as pilgrimages, binds piety to empire and casts sanctity upon endeavors whose most significant products were material.[16]

The work performed by Christianity in Bacon's text corresponds to its deployment in the New World within late Renaissance culture. But instead of a mandate for the colonization of bodies, Christianity becomes the code for an intellectual imperialism: a desire to take and absorb and control is legitimated by the sign of the cross. Hence, its presence at the close of the Bensalemites' first message to the Spanish sailors. Despite the peremptoriness of that message—the Spaniards profess themselves troubled because they are summarily warned off and forbidden to land—the sign of the cross was "a great rejoicing and as it were a certain presage of good" (3.130). The cruciform sign overrides all ambiguity. Like the vision of Constantine, its presence signifies ultimate favor, indeed, the power to conquer.

In Bensalem it is the phenomenal world of nature, rather than the natural world of savages, that stands as the object of conquest. The cross equates Bensalem's suavity and moral superiority with its domination of nature; it then follows as evidence of that superiority that their Christianity forecasts their scientific accomplishment. Such a basis of affirmation is circular: the conventional European definition of moral superiority presupposes Christianity, and thus Christianity cannot be adduced as support for a goodness that, when the sailors first read the communication, remains to be proven.[17] The text allows simultaneity to stand in for causality, and thus it argues for a natural philosophy endowed with conventional religious values by juxtaposing the two elements.

Such rifts can be seen only retrospectively; still, throughout *The New Atlantis* the pious philosophy of nature is an elusive production. At no time does this elusiveness become more apparent than when the Spaniards are told how the Bensalemites have come to Christianity independent of apostolic mission or latter-day conversion. This moment also provides evidence of the structures of governance controlling the text and of the inextricability of the spectacular and the authoritarian in the seventeenth century. When the governor of the Stranger's House relates the origins of Bensalemite religion, his utterance is highly marked with the signs of its own artificiality; to put it simply, he tells a tale. The coming of Christianity to Bensalem thus stands strongly differentiated from the other pieces of information imparted to the Spaniards: the governor's recitative has a formulaic beginning—"it came to pass" (3.137)—and a cadenced ending—"and thus was this land saved from infidelity" (3.139). All that lies between these two phrases stands in support of the mass conversion as narrative (and hence as cultural) exception by reminding the seamen again and again of its very constructedness. The scene of revelation is carefully set, and subsequent details are ordered to culminate in the surprise of the revelation itself. The Spaniards are thereby made witness to what is, crudely speaking, a dramatic event, whose denouement resolves the suspense created by the artifice of the narrative.

But the analogy to drama goes even deeper, for the scene of revelation itself contains a "spectacle." A "pillar of light," surmounted by the (transcendental) cross, appears at sea; the inhabitants of the coastal city Renfusa gather around, become its audience—but they cannot reach it in their boats, which "[stand] all about as in a theatre" (3.137). Nevertheless, one ship is allowed to approach, and it contains a Father of the House of Salomon, the philosophical institution that is the glory of Bensalem. When he rows to the pillar, he finds a watertight cedar chest that opens of its own accord. The chest is the engine of revelation, for it contains Scripture, other religious writings unknown to Europe, and a letter ordaining that the landing place of the ark would be granted salvation (3.138–139). The people of Renfusa see the action clearly, but they, like any group of spectators, like any audience at a play, are constrained from entering the stage, from crossing over the formal boundary that demarcates spectacle from experience, the seen from the lived-through.[18] The appearance of the pillar of light, which marks the discovery of the Scripture, defines the Renfusans as a chosen community, yet it excludes them from direct participation and apprehension; save that their knowledge is optical rather than auditory, their knowledge of revelation is but little different from the Spaniards'. The Renfusans are marginalized even as they are blessed.

The Salomonic Father, however, is not marginal to the spectacle but, rather, part of it; his escape from the general paralysis indicates such boundaries as those that separate audience from performance can be crossed as a prerogative of power-knowledge. The Father's actions valorize the inquisitional insofar as he sets forth to investigate what at first sight is but an exceptional phenomenon of nature. The text at this juncture proclaims the Father's philosophical college "the very eye of [the] kingdom" (3.137), and thereby banishes any presumed opposition between the scrutinizing practices of the Baconian scientist and the capacity for moral enlightenment. His pious initiative defines the centrality of the college to the spiritual life of Bensalem; for the moment, however, his singularity overrides his corporate identity. He alone of the crowd truly views the spectacle of the cross correctly, truly ascribes to it the proper divine significance in spite of its ambiguity. He alone appears to know that he participates in an act of divine revelation: "I do here acknowledge, and testify before this people, that the thing which we see before our eyes is thy Finger and a true Miracle" (3.137). He is central participant in the scene of revelation because he helps to produce it, because his presence at the event is part of its reason for being. By comparison, his countrymen are but mere consumers. As Stephen Orgel has similarly argued concerning the monarch of the Jacobean masque, the Father is uniquely capable of mobilizing the spectacle, because he alone provides it the principle of intelligibility.[19]

The collegian's exceptional percipience raises questions about the site of authority in Bacon's Atlantan society. If the Father presides at a moment of such fundamental importance, who can be set over him? But preliminary to this question is the issue of the spectacle's intelligibility: what cultural work does it perform, and for whom? At first reading, the spectacle seems principally to insist that Bensalem is exceptional precisely because its Christianity is revealed, immediate, rather than discursive and historical; this is certainly the point of the tale as told to the Spaniards. Yet the Salomonic Father prays to a "Lord God of Heaven and Earth" and knows the ends of miracles from books before he has ever set eyes on Scripture. The college of which he is a member is named after an Old Testament king at its founding, but that founding antedates the miracle at Renfusa. The Bensalemites do not live under the Old Covenant, for the Hebrews settled among them are classed with the Indians and the Persians; for all that, they seem to have prior knowledge of the basis of Christianity. Since More's foundational text had shown it possible to represent a society distinct from (if not alien to) received Christian doctrine, *The New Atlantis* is not forced into orthodoxy through the tacit coercion of precedent.

On the contrary. The scene of revelation is an act of mystification predicated upon the explanatory power of aesthetic modes in the Renaissance court. The dramatic markers invoked to set off revelation from the fabric of the quotidian underscore it as exceptional—but the nature of that exception derives just as much from the differing epistemological claims it makes as from any sense of the Bensalems' favored status with respect to the divine. Seen against the broader context of seventeenth-century culture, the tale of revelation responds to pervasive anxieties about the congruence of piety and the interrogation of nature, which amounted to human attempts to compass the province of the divine.[20] The Bensalemite tale does not validate interrogation by revisionary cultural labor, as when the emblem of Icarus began to be reglossed. Nor does it, as Galileo was to do in his *Letter to the Grand Duchess Christina*, attempt to establish a difference between what Scripture does and does not have authority over. The Renfusan set piece explains by declining to explain; instead, it asserts through the symbolic presence of its literary construction the transcendent legitimacy of what it represents, to beguile opposition by the operation of the aesthetic. As with Bacon's reference to *Amadis de Gaule,* real historical circumstance is domesticated by the shaping and ordering of literary form. The illusion of theatrical participation is sustained by immediate events in the narrative, since the governor who recounts the tale disappears after the performance to maintain its exceptional effect—and also to circumvent further inquiry. That the tale succeeds at its job of dispersing anxiety is borne out within *The New Atlantis:* the references to crosses and oblations, to piety and Christianity, which crowd the earlier pages of Bacon's text virtually disappear thereafter.

The scene at Renfusa is proclaimed a "true miracle" for the benefit of the populace; but when it is represented for the benefit of the Spanish seamen it becomes illusory, shifting, difficult to locate in any fixed system of meaning. Although Renfusan theological grace appears but seconded by its philosophical preeminence, it is that preeminence that underwrites its superiority to European culture. When it is compartmentalized in this way, Christianity becomes contingent rather than necessary. From there, it is the more readily recast as a free-floating legitimation of scientific authority, a potential accompaniment to an institutionalized meritocracy, rather than the ineluctable justification of kingship. That such a meritocracy, founded on the prerogatives of natural philosophy rather than on inherited divine right, could be thought into being is borne out by the representation of civil authority in *The New Atlantis.* In Bensalem, it is the College of the Six Days' Works, the institution whose Father figures so centrally at Renfusa, that provides both the political and the symbolic ground for New Atlantan society.

In striking contrast to its humanist pretext, *The New Atlantis* never elucidates its civil hierarchy, never gives articulation to its structure of power. Rather, the text uses the words "state" and "kingdom" interchangeably, although the two words summon up images of government that pull (and would come to pull the more) in divergent directions: one is corporate, if faceless, the other incorporated by a titled head, known to all. It is possible to read this silence about the authority at the heart of Bensalem as an affirmation of social and political conservatism, the more likely because the text that contains it is the product of a Jacobean bureaucrat in search of patronage for his philosophical program. In this case, the insertion of the Baconian agenda into an imagined social milieu much like that of early seventeenth-century England becomes a matter of tactics, of strategies of accommodation to the status quo, rather than a radical departure from received culture like the *Utopia*.

Yet Bacon's text also licenses such a departure. To be sure, Bensalemite society bears signs of congruence with its Jacobean correspondent, and the resemblance domesticates the Baconian program to the contours of seventeenth-century actuality. But to say as much is to suggest that *The New Atlantis* conforms to some, at least, of the expectations of utopian form, as I have already suggested. What is more remarkable is the manner in which the unarticulated, because unquestioned, can be transformed into the unnecessary. The structures of control emanating from the monarch presumed to govern Bensalemite society may as well not exist, since by the de facto weight of the narrative the text invests authority in the alternative of the scientific. In the space opened up by that alternative resides the possibility of reading otherwise, and hence of imagining a society not strictly tied to Stuart absolutism or to any other avatar of monarchy.

Authority is figured within *The New Atlantis* in ways whose complexities suggest the potential for new social formations. While the Spaniards meet several men of rank in the course of their visit, none can be clearly placed within a structure of governance; nor does the narrative accord weight to civil hierarchies. Instead, the role of authority is nearly always filled by the Fathers of the House of Salomon. Besides the twin displays of piety and acumen provided the sailors by the tale of conversion, there is the advent of the Father who takes over *The New Atlantis* by introducing the sailors to the works of his college, and by giving them permission to reveal what they have learned when they return. He is at once source and patron, only begetter of the utopian narrative; his disclosures concerning the work of the college, and his permission to make those disclosures public, license the Spaniards to speak of their unprecedented and novel experiences. The very mode of his arrival proclaims his importance to the text: he enters dressed in solemn black, bedecked with jewels, and carried by horses in a chariot without wheels and set with precious stones. He has many attendants, and all the citizenry turn out to greet him, as if he were a monarch on progress; as he passes, he blesses them silently. The pageantry makes clear that he draws all forms of authority to the House of Salomon either directly or indirectly. In *The New Atlantis* narrative authority bears the weight of public responsibility. In the process of representing the emergent structures of modern science as the basis of ideality, the text localizes those emergent structures in the Father and codes him as the de facto ruler of his land.

By indirection, the Father's narrative importance marginalizes the operations of kingship within Bensalem. Monarchy is in fact displaced to an antiquarian concern, since the only monarch of whom the Spaniards are told lived two millennia earlier. Although that monarch, Solamona, is profoundly venerated as the founder of both Bensalem and its college, he cannot be connected to either successor or descendant simultaneous with the arrival of the Spaniards. The history of his accomplishments bears a great resemblance to the myths of origin so useful to Tudor and Stuart rulers in legitimating rule through the invocation of impeccable (if fictional) antecedents: yet in *The New Atlantis* no evident textual ruler mobilizes such forces in the first place.[21]

There is, however, a clear address to Bacon's historical sovereign. The orthographic similarities between Solamona and Solomon demand that the mythic benevolent king of Bensalem be connected with James I, who was often figured as the Old Testament ruler, in Bacon's writings and elsewhere.[22] When viewed in light of James's well-known avoidance of the public manifestations of kingship, the Bensalemite hierarchy seems designed to provide a blessed blind.[23] It precisely avoids the precariousness of display unwittingly manifested in James's own *Basilikon Doron,* since in the Bensalemite model a king who never presents himself, and who in turn is never represented, cannot be held to account for misseeming. Only approved icons of kingship may be passed around, like the seals on the charters granted during the Feast of the Family; the body which provides the stamp of that image is withdrawn from circulation. Under this dispensation, the king is represented by his ceremonial and instrumental extensions, by the institutions and agencies whose very participation in the symbolic forms associated with kingship guarantee, so it would seem, the perdurability of kingship, the omnipresence of monarchical control.[24] The Bensalemite monarch's public and ceremonial functions are apparently displaced in this way, both onto the college, composed of one group of "Fathers," and also onto domestic life itself, where the father of every family reenacts the moment of coronation in his own domain on the Feast of the Family.[25]

But if the field is broadened, it is possible for this displaced authority to do other cultural work—for the occultation of any current ruler of Bensalem to be as significant within Baconian ideology as an obeisance in the direction of his Jacobean prototype. If, as Jonathan Goldberg has written in his account of Jacobean literary politics, "Sovereignty is a matter of sight," then *The New Atlantis* reconfigures the scopic regime of monarchy for its own purposes: by removing the king from the sight of his subjects, dispersing his gaze, and hence calling his existence into doubt, *The New Atlantis* inserts the philosopher of nature into the space of authority through the process of ocular reassignment.[26] Apart from the traces of the royal image that can be discerned on such ceremonial occasions as the Feast of the Family, the king's disappearance from the discursive field precisely coincides with the eminence of the House of Salomon, as the scene of revelation has indicated. These men are the "eyes" of their kingdom, the royal intelligencers Bacon advocates in *The Great Instauration,* whose very omnivoyance makes them as potentially destabilizing to monarchical hegemony as useful to its maintenance. In *The New Atlantis* the House is the "lanthorn" (3.145) of the realm—a source of visionary illumination, to be sure, but one which suggests the comparative obscurity of the king as well as the benightedness of the rest of the realm.

The need for vigilant observation is a persistent topos in Renaissance ideal societies; from the relentless communality of More to the monastical rigor manifested in Tommaso Campanella's *Città del Sole,* virtue is most reliable when cloistered within community. The Baconian text, however, takes the mutuality of supervision characteristic of More's text and places it at the control not of a moral hierarchy but of an overtly political one. The dismantling of horizontal relations, symbolized above all by the objectification of nature that Baconian science is predicated upon, has its necessary counterpart in this reaffirmation of social hierarchy on what is, after all, still alien ground. By the inevitable warrant of history—which is to say, under the conditions that determine the emergence of the modern ideology of science—the king as authorizing principle sits at the head of that hierarchy. But his presence cannot be detected easily, and the government seems to work efficiently without him.

The occulted, disembodied monarch of Bensalem can then be read in two ways: as imaginative reconfiguration of the Jacobean monarchy, and as model for a subsequent coexistence between scientific investigation and the dismantling of Stuart absolutism. The text forces recognition that the Bensalemite monarchy contains the possibility not just of its own effacement, but of its own erasure from the system of social and political order. The apparent vacancy at the monarchical center of *The New Atlantis* embeds within it an alternative text, susceptible of a different reading in light of the agenda of the English Revolution, and of social reformers during the period interested in the Baconian program.[27] The removal of natural philosophy from the possible disposal of royalist prerogative is sustained by the structure of *The New Atlantis,* with its kingship most potent in the mythology of the past, and with the hegemonic presuppositions of the House of Salomon obscured by the scattered signs of conventional authority.

It would be facile to argue that Bacon's text enabled the English to unthink kingship, to posit it, in short, as a prelude to regicide.

But *The New Atlantis,* influential as it was in the period after its publication, must be seen as a constituent of post-Jacobean culture whose productions were as various—and yet as interconnected—as the Invisible College and the Commonwealth. And if Bacon's text operates as a mechanism for the ideologizing of its audience, its philosophical agenda cannot be separated from its representation of power. It then remains to be seen how *The New Atlantis* introduces such process into the fixity of its utopian form, how it reforms humanist literary politics into one of the mechanisms of modernity.

In whatever way *The New Atlantis* produces revolutionary readings through its abeyant dismantling of monarchical hegemony, the issue of authority within and without the text, and of the means by which the text opens up the space for difference, cannot be resolved solely by addressing the thematized structures of governance bequeathed by English history. If, as I have suggested, the utopia is a form which plays upon, yet neutralizes, the pertinent conditions of its historical emergence, it then follows that treatment of the form itself is a matter of authority, a testament to the pervasiveness of humanist texts as a cultural model and a point of difference for an emergent discourse.[28]

Still, *The New Atlantis* is as much a reassignment of More's utopian form, and of its relationship to the culture out of which it emanates, as it is a testament to the availability of humanist texts. The closeness of many Bensalemite social practices to those inscribed within Jacobean society, even as regards the problematic representation of monarchy, runs counter to the distance the second book of the *Utopia* maintains from the concrete historical circumstances presented in the first book. What I have earlier termed utopian equipoise in More is given up by Bacon in his writing of differentiation, which makes the text at once the colonizer of preexistent form and the producer of ideological novelty—the regime of science—from within that form. In brief, what was once humanist paradox, imbued with the stuff of Renaissance courtly politics yet impracticable as radical agenda, has in Bacon's text become something very close to propaganda.

The discursive colonialism of *The New Atlantis* toward its humanist prototype is an important move in the all-but-silent prehistory of
disciplinary formation: the emergence of "modern science" and the
concomitant ideologies of culture that deem the humanities the antithesis of science depend upon and emanate from this othering of literary
representation by the scientific in the late Renaissance.[29] The pertinence of colonization as a model for this process—which is to say the
pertinence of such historically specific structures of domination and
alienation—is determined by the moment of emergence. *The New Atlantis* was written, and indeed could only have been written, at a particular juncture in Jacobean culture: at that juncture, the fact of the New
World was no longer novel, but the issue of how to control it, imagina-
tively as well as materially, had become critical. The sheer amount of
writing generated about the Virginia Company's ventures not only
attests to the glamour of novelty, but also evinces inadequate discursive
mastery of the new. Thus, Tobias Matthew, archbishop of York, wrote
in 1609 to the earl of Somerset: "Of Virginia there are so many trac-
tates, divine, human, historical, political, or call them as you please, as
no further intelligence I dare desire."[30]

Obviously, the New World informs much English Renaissance
writing besides Bacon's—in More's *Utopia,* for instance, its most proximate text. But in More's realm the comprehensiveness of description is
a measure not of the need to master and assimilate the unprecedented
experience of the Americas, but of the tightness of social control within
the geography of the imaginary. The text presents a wholly realized
society to make the invariance of that society concrete, to reify the
stasis that is a precondition of Utopian ideality; thus control is writ
large across Utopian life. For More, the New World is not a site for
further exploration, but an informing presence whose newness "explains" the desire to create society anew, an impetus to formal and
cultural innovation.

Bacon's advocacy of a revisionary philosophy of nature, in contrast,
shifts the focus of his utopian text from the formalized study of human
society to the systematized knowledge of an alien natural world. Such a

shift in positioning—which is, perhaps more importantly, a shift in object, in semiosis—brings with it a dynamism not operant in More's text. There, the reconstituted, completed social institutions of Utopia focus on the processing of a unanimous subjectivity, bespeaking the agenda of humanism even as those institutions appear to hark back to medieval monasticism. The Baconian text, however, stages inquisitional structures almost proleptically, as advance notice of a system whose material emergence its representational practices then make possible. In the way that *The New Atlantis* emphatically polarizes the flow of knowledge typical of colonialism, from the inhabitants to the seamen, from those granted the power of natural philosophy to those comparatively disempowered, it turns the disempowered into objects of scrutiny themselves, into displaced versions of the nature presumed (and narrated) to be the primary site of investigation. Insofar as a reader may be positioned likewise, *The New Atlantis* reconfigures the bemused idyll of More's *Utopia* into a narrative that privileges novelty, process, and change.

This bend away from utopian convention is coterminous with an assimilation of narrative practice to the emergent structure of empiricism, an assimilation signaled at the outset. *The New Atlantis* begins with lost sailors, desperate not because of storms but because of a strong wind which has carried them off course, leaving them in becalmed waters. Then the Spaniards spot "thick clouds" which put them "in some hope of land" (3.129); shortly thereafter, they arrive at Bensalem, and their education commences. Such a beginning may well be termed uneventful, since it apparently seeks to avoid the impetus to travel narrative lent by storm and destruction, or for that matter the potential importance of a thematized break to signal the beginning of a text.[31] Its eventuality, however, is not to be located in prior literary practice, but in the historical space that connects the New World with the New Science. The clouds the sailors see, and take as a sign that they are near land, are habituated observations that belong with the necessary skills of the seaman, and here provide textual emphasis for the correspondence between exploration and investigation. The sailors' acquaintance with natural phenomena effectively prefigures them as agents of the philosophical inquiry carried out by the Fathers of the House of Salomon. Accordingly, Columbus writes in 1501: "the life of sailors . . . leads those who follow it to wish to know the secrets of this world."[32]

Within *The New Atlantis* such "secrets" are coded as a sign of initiation. While the Spaniards demonstrate their competence in the rudimentary decipherment of clouds, it is not until they submit themselves to a superior (because systematized) order of knowledge that they gain access to what the world, unless through the college, keeps to itself. The means of access put a spin on the epistemology of sight, enlist it to serve the emergent ideology of science: what is known through seeing is reconfigured as more real and replaces Hythlodaean tale-telling with an optical warrant whose signs are themselves available to be seen, read, and hence believed. *The New Atlantis* is advanced—propagated—by the agency of vision, just as authority within the kingdom of Bensalem inheres in its collective "eyes." But there is more than analogy at work here: the dispersal of kingly authority initiated by the text's bequeathing his specular prerogative to Salomon's House has its sign in the narrative obsession with speculation about nature as spectacle, with knowledge as a form of display. Within the ideology of scientific modernity, the eye is the sign of knowledge. It follows that a text designed to naturalize that knowledge would depend upon what is seen as a form of self-legitimation. Observations construct the world of the text as the enactment of the inquiry it espouses: insofar as *The New Atlantis* teases out its own agenda, it is through the accumulation of scrupulous detail. These details say less about the utopian valence of Bensalem than about the way that it comes to be known to the Spaniards—and by extension the way that the text itself is directed to the realm of material, historical practice.

This positing of the visual as the arbiter of both representation and practice with *The New Atlantis* accounts for the multitude of parenthetical phrases within the text. Sometimes, the phrases bracketed off from the main narrative in this way amplify upon an observation; they interrupt its forward impetus in order to scrutinize an object: "he drew forth a little scroll of parchment (somewhat yellower than our parchment, and shining like the leaves of writing tablets, but otherwise soft and flexible)" (3.130). At others, however, the enclosed elements call attention to the tenuous phenomenological basis of narrative convention. The parenthesized information betrays the assumptions of omniscience as it focuses on the perceived, and therefore tentatively known, significance of event or act: "there came towards us a person (as it seemed) of place" (3.131); "whereupon one of those that were with him, being (as it seemed) a notary" (3.131); "in his hand a fruit of that country. He used it (as it seemeth) for a preservative against infection" (3.132): "he smiling said, 'He must not be twice paid for one labour': meaning (as I take it) that he had salary sufficient of the state for his service. For (as I after learned) they call an officer that taketh rewards, *twice paid*" (3.132).[33]

This anxious display of warrant makes authority within the text an index of emergence. The parenthetical eruptions not only put under suspicion the local issue of representation as omniscience, of the literary as epistemology; through their obsessive interest in establishing an ocular, verisimilar basis for the text's knowing what it does, they also signal the shift away from literature as hegemonic practice in Renaissance culture. By foregrounding a pragmatics of narrative information, they question the perfection bestowed by Sir Philip Sidney on poetry as a function of its role in the construction of humanist dominance—a point that can be extended to the ideality of utopian formations.[34] Instead of either gilded improbabilities or internally conflicted ideals, *The New Atlantis* presents narrative as potential fact, and the work of reading the exercise of an acumen newly harnessed to the production of scientific knowledge. Bacon's text needs to be processed as the collegians process the material of nature: to be scrutinized, sifted, and disposed of according to a nascent taxonomic hierarchy, where only that which can be accounted for by means of physical apprehension or its narrative embodiment is accorded significance. The interpretation of any such data within the utopian frame of the text is steadfastly resisted, for there is always more to be known, and that superfluity denies formal closure. Such open-endedness has been deemed characteristic of modern scientific cultures.[35] Here, it lends Bacon's text the dynamism that I have mentioned. But that dynamism is antithetical to human-ism's utopian forms, which impose the fiction of coherence on the land-scape of ideality, much like the coherence posited upon the natural world by hypothetical formulations in contemporary Continental sci-ence. In a narrative constructed by the all-consuming eye, however— and here I enlarge the term "narrative" to include Bacon's nonutopian texts—there is no room for unifying fictions, no cessation of the activity of processing, no apparent imposition of hypothetical form on the body of nature deconstituted by taxonomizing scrutiny.

The New Atlantis is literally an incomplete text: the last page con-cludes with the editorial note "The rest was not perfected," introduced at the place in the text after the Spaniards have been granted per-mission to disseminate information about Bensalem as a gesture of goodwill. There is a correspondence between this gesture toward dis-semination and the cessation of narrative impetus: the fractional state of the text can be assimilated to the way the text dictates practice. To argue such a correspondence may appear to run the risk of overstaking a claim on the accidental. However, when all of Bacon's texts are to one degree or another "unperfected"—from the twice-revised and twice-augmented *Essays* to *The Great Instauration* and all the components nested within it—it becomes fair to consider the textual and ideologi-cal functions of imperfection in the Baconian program.

On one level, of course, the incompleteness of any Baconian text is a testament to its messianism. As the frontispiece of *The Great Instaura-tion* evinces, the recasting of knowledge undertaken by Bacon must be understood through structures of liminality that represent the moment of change—the transition from old to new—rather than its accomplish-ment. But this claim, in turn, demands to be inserted into the specific textual framework of *The New Atlantis*. "Perfection," properly, is com-pletion; it also signifies that which is perfect, the ideal—the utopian. *The New Atlantis* cannot be "complete" in this specifically utopian, specifically textual sense, since it generates a model of continuous ac-tivity that converges on the possibility of progress. It becomes a prime component of an accomplished and institutionalized scientific ideology, whose domination of the natural world the text seeks to authorize, even to inaugurate, by the means of its narration. The admonition to publicity on the last page of the text indicates that the work of science is durably configured by the Baconian text as open-ended, as an ever-deepening knowledge of the secrets of this world to be shared among initiates. At some point the eye will have seen enough because the eye will have seen all: but the possibility of such transparency depends upon reiteration and repeatability, upon the sense of a shared and duplicated agenda that still characterizes the production of modern science. Like the endless romance of the New World in early Stuart England, (post)Baconian science is always, it seems, in need of inscription.

What is the function of gender in this inscription? To echo the question posed by Luce Irigaray's essay, "Is the subject of Baconian science sexed?"[36] As that punning title suggests, Irigaray connects up science as a discipline with an apposite object of study, and science as an agent of subject-formation: if women are not addressed as subjects by and within scientific practice, Irigaray argues, it follows that the "subjects" taken up by science appear only to be gendered masculine.

Certainly, few overt moments of oppositional gendering occur in the utopian dynamic of *The New Atlantis*. Unlike either *The Tempest* or Hall's *Mundus Alter et Idem*, Bacon's text, in its colonial aspirations, seems distant from a ready accommodation of femininity and nature. Its subjects may—indeed, must, given the intertextual pressure of seafaring narratives—be gendered male; further, the coincidence of masculinity and control are to be seen throughout the Baconian program, most specifically in the particular domestication of natural philosophy to the dynastic exigencies of the Jacobean court.[37] Nevertheless, the instance of gendering most pertinent here depends upon a conceptually prior and deprecatory sexing of (literary) textuality, a space where poetic icons are divested of cultural authority and set over against it through gender binaries. *The Advancement of Learning*, for instance, is itself an agent of differentiation, positioning itself within humanism in its referential structure while seeking to stabilize the runaway signification it finds there. Even as the classical learning of humanism is adduced for the purposes of argument, it is indicated as evidence and marked as an occasion of potentially dangerous (because pleasurable) excess. That pleasure depends on linguistic fetishism, the phantasmatic substitution of a part for a whole, a false image for a true thing. All artificial signs become equivalently simulacral, whether words or paintings or statues: "It seems to me that Pygmalion's frenzy is a good emblem or portraiture of this vanity: for words are but the images of matter; and except they have life of reason and invention, to fall in love with them is all one as to fall in love with a picture" (3.284).

As the portrait of erotic transport makes clear, Pygmalion's desire codes the disruptive power of "images" through a feminine body, where a neo-Platonic correspondence between the good and the beautiful falls victim to a play of false surfaces. The work of metaphorizing contains the prevailing cultural force of language within social relations and collapses the hegemony of fictions into a private affair. In *The Advancement of Learning,* this mythic frenzy, a fruitless love for an inert and unproductive form, stands as emblematic for the excess investment in language that a direct correspondence between words and things means to correct. It stands to reason, then, that the "matter" to be rendered productive is less the body of nature—especially given the notorious fruitlessness of the Baconian investigations recorded in, for example, *Sylva Sylvarum*—than the matter of words, the humanist corpus.[38]

To bring together *The New Atlantis* and its humanist Utopian pre-text with the overt figuration of gender, consider the marriage inspection, which each text stages and which Bacon's text seems to take over nearly wholesale and deadpan from More. Joabin, a Jew who lives in Renfusa, is sent to supply the Spanish sailors with the information about Bensalemite culture, including their prenuptial practices. By way of establishing the superiority of New Atlantans, he offers a gloss on the *Utopia*'s sly parody of marriage contracts, which like horse-buying, necessitate a thorough inspection of the goods before purchase:

I have read in a book of one of your men, of a Feigned Commonwealth, where the married couple, before they contract, to see one another naked. This [the

Bensalemites] dislike, for they think it a scorn to give a refusal after so familiar knowledge; but because of the many defects in men's and women's bodies, they have a more civil way, for they have near every town a couple of pools (which they call *Adam's and Eve's pools*) where it is permitted to one of the friends of the man, and another of the friends of the woman, to see them severally bathe naked. (3.150)

As though the polyvalence of More's text were a form of provocation like the one that so enraptures Pygmalion, Bacon restricts its playfulness about the appeal of bodily surfaces. Instead, Bacon's text renders More's play *productive*. The *Utopia* offers a useful model upon which to build a rational theory of choice in marriage, needing emendation only to account for the consequences of "familiar knowledge" through the interposition of trusted intermediaries and through a multiplication of the bodies available to be gazed upon. In the process of correcting More's parodic depiction of sexual inspection, *The New Atlantis* offers a further means of legitimating empiricism even within the erotic—or perhaps of de-eroticizing that erotic, the better to recirculate its energy in the emergent program of the scientific.

To call such a pool after Adam and Eve is to insist on the innocence of the knowledge gained there. The visual apprehension whose safety and discretion are guaranteed by secondhand sight has no idiosyncratic, and hence no libidinal, component. Rather, a communal standard of attractiveness and marriageability seems to obtain, a general, even generic sense of the appeal of the unblemished body. In contrast, the pleasure Milton's Eve and (especially) Adam have in the sight of one another's bodies is rather more complicated—as is the role of vision throughout *Paradise Lost*. But however perfect Adam and Eve are in their appeal, they are no match for the prosthetic body that haunts Milton's poem. The quasi-satanic, quasi-theoretical eye of modernity watches them move through Eden; it watches them fall into history and temporality. Indeed, the historicizing gaze that haunts Milton's text interrogates not just their every move, but their adequacy as the avatars that late Renaissance humanism offers for universal truths.

4

The Prosthetic Milton; Or, the Telescope

and the Humanist Corpus

Eden is garden, not island, heavenly pendant, not a jewel scattered on the blue. But the tropes of removal and singularity that in *The Tempest* made the island a privileged locus for reading the emergence of modern subjectivity—that made it, in other words, the phantasmatic topography of modern times—have a clear application to the privileged locus within Milton's text. When paradise is designated, as Milton's Eden is, to be God's "new world" (2.403), its placement within the thematics of colonialist novelty marks it as an alternative topography, a stage for the cultural problematic traveling along with the formation of the new.[1]

Conjoining the New World with the New Science as the semiosis of modernity has the suppression of embodiment as a condition of the subjectivity it models, as the analyses of *The Tempest* and *The New Atlantis* have suggested. The idealized (island) locale of Shakespeare's play serves as Prospero's dynastic laboratory, but his appropriations of it—continuous, of course, with his appropriation of Others for his work—speak to the island's more general utility in the devising of new social formations. *The Tempest*'s protoscientific machinations install two modern registers, the domination of nature and the domination of the other, in a hierarchical partition of the (male) mind and body, while leaving the feminine as the ground of possibility for all such discursive changes.

These analytical counters still circulate in Milton's exceptional space: indeed, they are given point by the poem's assignation both of hierarchy to gender and of gender to work.[2] But Eve insures that the feminine is inevitably foregrounded by plot in Milton's Edenic improvisation on alien geography. The division of labor between mind and body apparent in *The Tempest,* which in that text maps onto the colonial and familial hierarchies it naturalizes, draws the sexes into the fray overtly in *Paradise Lost.* Although Adam and Eve together work the garden as equal partners, more telling labors are pursued by each without benefit of conjugal partnership. These, like Eve's cooking for their angelic guest, establish the body as the focus of her attention and (en)gender the nascent division of the corporeal from the intellectual. Certainly the crucial event of the poem, the Fall, pictures Eve as "separate" (9.424), going about the work of sin, which as eating, as appetite, is also the work of the body. But an earlier, "innocent" moment in the poem, where Adam and Raphael share in the rarefied cosmological issues attendant upon the emergence of scientific discourse in the seventeenth century, may be just as revealing. When it comes to studying the heavens, Adam inquires while Eve retires.[3]

To bring together the manifold categories of analysis that Milton's epic demands, I want to focus attention on a comparatively incidental aspect of *Paradise Lost:* the telescope, whose textual presence occasionally interrupts the poem's unfolding of events before the commencement of history. As an apparatus that signifies the "New Science" of the seventeenth century, and a technology that makes "new worlds" available for inspection, the telescope seems a useful index of the transfer of cultural authority from humanism's printed texts to colonialism's and science's natural ones. Further, as an instrument of formal and ideological reproduction it rounds on some of the issues concerning the body, both in history and through technology, raised by my initial examination of Cindy Sherman. How does Milton's poem speak to an emergent relationship between the body and its projections? What happens to that body and its gender (if indeed it is not proleptic to separate them) when the telescope enters and shapes the scene? How does the telescope designate the Miltonic landscape as "space" despite itself, and so betray the colonization of the humanist text by the discourse of scientific modernity?

The cultural problematic subtended by the telescope demands reframing the division between subject and object to make it a matter of history—in this case, to represent it indirectly as a split in the early modern subject, between the (dis)embodied mind's eye and the material component brought in to supplement it. Generally, the telescope of my chapter title has been regarded as "theory materialized," in Gaston Bachelard's phrase: an emergent theory of astronomy that, in contemplating distance, makes distance from the corporeal observer a primary constituent of its system of meaning.[4] Certainly much could be made of how the references to astronomy in *Paradise Lost* figure distance in comparison with the long view of the divine, and thus of how they augment the perspective of theology with that of science. Marjorie Hope Nicolson, for instance, has generally characterized the epic as "the first modern cosmic poem," noting that the action has "interstellar space" as its backdrop.[5] More recently, Harinder Singh Marjara argues that *Paradise Lost* be read as an original variation on emergent scientific themes available to Milton, who "erects a much broader and grander vision of the universe than the scientists, since he makes his vision relevant to his larger poetic purpose."[6]

But adopting this general line of analysis means positioning the telescope as a sign of spatial removal, and so as the sign of an astronomical theory conceived apart from the contingencies and contradictions of its early formation. Perhaps more important, it also would mean privileging theory as a signifying practice over the technology that, often, makes theory possible. Instead, I want to cross-examine Bachelard's granting priority to formulaic abstraction over materiality, a concession that effectively places scientific discourse in the realm of the eternal and absolute. While the formation of such an eternity as a particular alternative to Milton's is one destination of this chapter, the history of that formation is another. Furthermore, I also want to claim that something odd happens to embodiment—human materiality—in the presence of this alternative (and potentially emulous) form of the material. In short, rather than collapse history into theory, I propose to make of the "Optic Tube" in *Paradise Lost* a register of the historical—a theoretical history, of course, which thus signifies a distance as much discursive or ideological, as physical or temporal.

It might be better, therefore, to shift the emphasis away from the act of knowing at a distance that the telescope models through a system of optical representation to the subject-body that is the point of application for such devices. Which is to say, the body both natural and artificial, as "created" and as altered, supplemented, discernibly worked upon. Both such bodies, of course, are bodies in and of culture. The distinction I intend follows instead upon the lines of inquiry established by Michel Foucault and Thomas Laqueur.[7] It assumes, in effect, that modern social formations, particularly those associated with scientific power-knowledge, bring into being a subject-body hypothesized as essentially different from that produced within the classical, and even early Christian, "pastoral of the flesh." To speak cannily (and punningly?) of the premodern experience of corporeality as a "pastoral," as Foucault does, suggests that the body is always itself a *fantasy* of nature under the care of a divine agent.[8] But it is important to stress that it is a nostalgic fantasy. Modern technologies of knowledge produce a clinical, truth-saturated nature when the body is probed, shaped, disclosed; unlike this empiricist nature, the pastoral nature in which the premodern, "natural" body is involved is prior in a very restricted sense—dependent, perhaps, upon the very fantasy of priority, and hence of a specifically Christian theology of loss.[9] By contrast, what I have called the artificial body—part subject, part object—may well be the first term in the series that ends with Donna Haraway's cyborg.[10]

The particular spin I have given to the word "pastoral"—which in Foucault is more particularly confined to clerical stewardship—makes clear that the garden is the site where this line of argument intersects with *Paradise Lost.* In Milton's epic the shift from natural body to artificial one coincides with the central drama of the poem: the body that falls and that (perhaps) occludes its weakness, repairs the defects of the Fall by recourse to the prosthetic. That repair is more Baconian than Scriptural, seeking consolation for the loss of paradise in the scrutiny of nature; for that reason, it does not constitute part of the conscious agenda of the epic. Because most of the poem attempts to render the prelapsarian cosmos present for moral inspection, *Paradise Lost* is more overtly concerned with the first category of embodiment. But the occasional play of optical devices in the text presents the alternative, the artificial body of the scientific project, as an inadvertent critique of that naively worn flesh. It is not a moral critique of display, however, which is to say not one confined to the thematics of the Fall and of the shame of nakedness. Rather, the telescope is a sign of discursive adjudication, which might anachronistically (and crudely) be deemed the judgment of scientific fact on theologically-inflected humanist fiction. Or a contest to determine what counts as a way to recount, produce universal truths—the absolute modern space from which Bachelard's theory emanates, and which in turn validates the authority of the theoretical.

Clearly, this critique, as I have called it, is not dictated by the plot of the epic, what it takes from Genesis and the hexameral tradition, although it does depend on the poem's attempts to tell the story of "man." Instead, it is produced by and within an ideological shift contemporary with the poem and that ultimately produces recognizably modern taxonomies of writing.[11] As an incursion from a different (and potentially oppositional) realm of textuality, the telescope, and the observer to which it is conjoined, render the frail flesh of our poetic antecedents a matter of detached speculation.

Let me follow up on these hints by making a textual problem of the references to be examined. It is fastidious and anachronistic to suggest that the telescope seems out of place in Milton's epic poem; but some claim of the sort undergirds the present analysis. Paying attention to a "naive" response that would wonder at the decorum of the references to astronomy, Galileo, and the telescope seems useful, notwithstanding how easy it would be to explain such wonder away. In other words, something worth fretting over is going on here. As other critics have done, I want to read this material, which the chapter title has designated a prosthesis, as a privileged signifier in the text.[12]

It is widely accepted that Milton visited Galileo in 1639 during Galileo's confinement at the end of his life, and that Milton was well acquainted both with the astronomer's views and their suppression by Catholic authority.[13] For John Guillory, the poem's references to the astronomer represent "the return of the repressed in history," a history understood as biographical and exemplary.[14] In positing a phantasmatic identification between the postrevolutionary poet and the astronomer, Guillory suggests that the proper name of Galileo be read as a "cryptic self-portrait" (161) of an author whose failed politics had led to a similar suppression and marginalization.

The name Galileo—which does not accompany all references to optic glasses in *Paradise Lost*—may well function as the covert sign of Miltonic identification, of the career of an exceptional man of letters caught in historical catastrophe through a problem in vision. But implicit in the very displacement that Guillory argues for is a parallel and received displacement of discursive authority—from the field of humanist poetry to that of astronomy. Elsewhere, Guillory has suggested that to read for gender, for example, is "to discover behind the struggle of Milton's fictional agents the giant forms of discourses in conflict."[15] This apt description of the epic's investment in history demands extension to its references to Galileo. Potentially, he is another such "fictional agent," albeit one whose range is restricted, compared to the universality of Adam and Eve. The proper name of the astronomer, and the uses to which it is put in determining the technological body, become signs of an anxiety about the eclipse of the humanist project by the emergence of an alternative ideology of the scientific. Simply put, the imperiling of the astronomer, which Guillory identifies with Milton and the consequences of his revolutionary politics, unconceals a moment of cultural precariousness—not solely of the particular writer John Milton, but of the late humanist producer of literature at a moment of discursive instability.

The telescope, like the name Galileo itself, registers the gap between the script of the poem, the metaphysics of origins, and the emergence of modern structures of feeling, between universalizing theology and contingent history. In fact, the topicality of such references may very well be the point of entry for a discussion of competing discursive formations. References to the telescope register a gap in the poem between the represented absolutism of a prelapsarian past and the infiltration of the topical. Thus the telescope makes a cultural problem out of what it sees—and, maybe, out of the fact that it sees at all.

In what sense can Milton's figuration of the telescope be established as "prosthetic"? The best place to begin the analysis is with the
passage in Book 5 that describes Raphael's departure from heaven to
"converse with Adam" at God's behest. As he descends, the angel's first
sight of earth introduces this simile:

> . . . As when by night the Glass
> Of Galileo, less assur'd, observes
> Imagin'd Lands and Regions in the Moon . . .
>
> (5.261–263)

Nothing here is overtly obscure: when editors bother to gloss this
passage at all, their laconic response goes something like: " 'Glass/Of
Galileo,' that is, the telescope."[16] Although the ultimate correctness of
the gloss is undeniable, it seems useful, if perverse, to tease out what
happens when "the Glass of Galileo" is not automatically translated
into the telescope, when the instrument is not yet considered to be
formally alienated from its inventor, not yet the accomplished technological object used within astronomical practice.

In such an instance of willed ignorance, what connects Galileo and
"his" glass? If the relation between them is taken to be one of possession or invention, then the case is still simple: the "of" is proprietary.
But what if the obvious reading is suspended, and "of" betokens a
partitive genitive, a designation for the part of a whole, as in the phrase
"the fingers of one's hand"? This reading then unconceals a ligature: the
preposition yields a glimpse of an apparition in which the authorizing
subject-body can be seen to extend itself into the apparatus, reconstitute an originary coextension with it. Here, the glass functions met-
onymically as a technological substitution for the eye into which a
surveying subject may be compacted. Who is the subject, after all, who
is there to "observe" the "Imagined" lunar plains? Presumably, it is
Galileo—but only as an attenuated consequence of the partitive genitive. If a more conventional reading of the preposition is taken, however, even that ghostly observer disappears, subsumed into the troubled
observations of a disarticulated, "less assured" lens.

This elusive reading, like the illusionary subject who wields the instrument, has been cut out by the normative construal of the preposition. If the glass is taken to constitute part of Galileo—even if "only" at the level of signification—the poem, however fleetingly, presents a particular representation of the modern body, beginning to be produced by what has been institutionalized as scientific practice. Primarily, it is a body which, like that of the illusionary observer, is about to be banished from the site of its production, one that is dependent on its telescopic armature to betray any trace of the observer's presence. Indeed, that armature supplants the observer as much as it supplements him: his eye (a more conventional partition of the body) has become "the glass." Construed in this way, the phrase "the glass of Galileo" refuses to be collapsed into an easy reference to an early modern scientist who uses an investigative tool separable from him in the service of an objective truth. Instead of that anachronism, the phrase offers a shifting sense of the relation between technology and the formation of the sovereign subject of science, a subject split from his body.

This less assured "glass," and the disembodied observer it subtends, provide the most resonant instance of the telescope as prosthesis in *Paradise Lost.* Yet even when the syntax makes clearer the physical distinctness of the instrument and author, the telescope is always connected to its putative maker, Galileo, the "Tuscan artist" (1.288), or the "astronomer" (3.589).[17] Given the physical conjuncture these instances suggest, the simple identification of "glass" as "telescope" elides a historical process, a process of sorting out the difference between (among other things) subject and object that is the work of an emergent scientific ideology.[18] In this precise sense there are no references to the telescope as such in *Paradise Lost.*[19] The "Optic Glass" whose glimpse of the moon is compared to Satan's shield in Book 1 (286–291); and the "glaz'd Optic Tube" which reveals the demoniac sunspot in Book 3 (588–590) differ from the telescope not only in taxonomic nicety—what they are called—but in how they seem to work. In none of these cases is the instrument presented whole, formalized, stable by and in name; instead, its utility as an apparatus of enhanced visual perception resides now in its lenses, now in its shape.

Even when Milton designates the optic tube a telescope, as he does in *Paradise Regained,* the text still makes a puzzle of the apparatus: its proper name, its function, indeed, even its presence—whether there is a telescope there at all—are all put into question. When Christ is lofted on high to view the earthly empire of Rome, the force that renders the panorama "presented to his eyes" so accurately is introduced as a matter of some doubt: whether "strange Parallax or Optic skill / Of vision multiplied through air, or glass / Of telescope, were curious to inquire" (4.40–42). The vision of earthly temptation that Satan lays out for Christ to view might be at the behest of enhanced but unaided sight, some nebulous quality deemed "Optic skill"; it might also be the Devil's work—or Galileo's. Although Satan ascribes the vision to the disposition of his "Airy Microscope" [*sic*] (4.57), his naming the apparatus does not disperse the prior uncertainty that the text has allowed to creep in. Quite the contrary. For one, Satan's claim to possess such an apparatus cannot be verified apart from his claim to access. Does he brandish it, compel Christ to look through it? As the subsequent reading of Galileo's *Dialogue on the Two Chief World Systems* makes clear, perhaps no one may be seen to look through a telescope, and presumably Christ least of all.

Then there is the question raised by the nomenclature itself: when, if ever, is a microscope a telescope? The two instruments are clearly distinguishable in the science that claims them and in the scale of their representations; from this vantage point, indeed, they seem to have been distinct when the technology of each was developed.[20] In *Paradise Regained* this blurred identification can certainly be read metaphysically. The hyper-enlargement of the minute that characterizes microscopy is, from a spiritual standpoint, precisely analogous to the inflated claims made by Satan: hence this reduction by instrumentation trivializes the devil's offer of glory. However, any such redemption of the poem from its confusion about the instrument of vision merely displaces that confusion onto the object of sight, by rendering it unstable, larger—or smaller—than it "is." In any case, in *Paradise Regained* as in *Paradise Lost,* the multiplicity of terms and causes around the telescope is destabilizing: it makes impossible a notion of instrumentation as fixed, as a neutral conduit for previously constituted visual phenomena to be examined with dispassion.

As this argument suggests, Miltonic glasses and optic tubes are not readily assimilated to modern apparatuses of visual investigation, however much in the end they can readily be glossed in *Paradise Lost* as contiguous with them. Nor is the point about the telescope here purely Foucauldian. I am not primarily interested in the emergence of new discursive objects or, for that matter, the docile bodies produced by technologies of surveillance—except where the disciplinary force of the telescope is just that, a matter of discipline in our contemporary sense, of field and of institutionally dominant formations.[21] Rather, in examining how the prosthetized observer of astronomical practice hovers at the margins of *Paradise Lost,* I am also claiming his augmented body as a counterpoint to the fallible subject of the Garden of Eden. And that subject, at its most fallible, is therefore most likely to be feminine. In the distance that separates the telescope from the optic glass there is space to consider how these references inflect the universal script that *Paradise Lost* aims to be.

One can read the telescopic moments as irruptions of contemporary history into *Paradise Lost,* as I have suggested. Drafted as a space before history as such and as a lost alternative to it, Milton's Eden still betrays the seventeenth-century formations that it partly constitutes and with whose contradictory positions it is inscribed. The nostalgia for green perfection whose ultimate (and dilated) referent is the lost Judeo-Christian garden, as Raymond Williams noted some time ago, is never so simple as a longing for escape from the world of human manufacture into the blank absolute of the natural, the ultimate register of divine plenitude.[22] Yet the presumed universality of such a desire touches on the specific work of Milton's quasi-pastoral, and may be partial proof of its success in literary history.

The "universality" to which I have been referring is not only doctrinal or nostalgic; it is also a function of humanist exemplarity, as the work of Timothy Hampton and Victoria Kahn can be read to suggest.[23] As a document of, and belated testament to, Renaissance humanism, *Paradise Lost* installs the truths, transnational and transhistorical in pretense, that travel along with Christianity. But it also exemplifies how "literary" universality comes to be seen as separable from religious doctrine. The poem's theological grandeur, its cosmic map, and the general dicta that attend its magisterial unfolding of a spatialized providential history appear paradigmatic for Renaissance humanistic ideology, perhaps even more than for modern science. As I argued in chapter 1, the moral exemplarity informing humanist philological practices makes it possible for contemporary male readers to compare themselves to, and map themselves upon, the textual figurations of antiquity. In deeming Renaissance humanism an ideology, therefore, I aim to call attention to the way it constructs its subjects through an imaginary identification with the textual models provided by classical texts. The identification becomes a means to produce dominant formations and, ultimately, a pedagogical practice to reproduce them. As Victoria Kahn has observed, reading classical texts as *exempla* effaces the particularity that makes it possible for nonidentity to signify: "[Humanists] forget difference precisely to the extent that they are interested in the ethical exemplarity of classical texts, to the extent that they conceive of reading as a form of practical reason" (468). Put bluntly, the imitation of reason produced within humanism insists at its core that the past is always pertinent because men everywhere are always the same.

The humanist effacement of difference, however, is not solely a classically inflected program of ratiocination. Rather, it is a politics of reading, the ideology of an international *res publica litterarum,* whose material practices are first dispersed in an era of transatlantic and transglobal exploration and expansion. Humanist textual productions can thus provide a script for the symbolic violence that informs the emergence of modernity, as Stephanie Jed's work can be used to argue.[24] Jed shows how the classical legend of Lucretia—the raped and self-incriminating suicide whose death led to the founding of Rome—was read and reread as a foundational text for Italian city-states. The very utility of the story redeems it from particularity: it neutralizes violence against women by turning such offenses as Tarquin's rape into the lamentable but civically productive condition for the emergence of the modern republic. As a result, Jed's study provides a way of bringing together a political culture that focuses on the good of the state with the dissemination of humanist exemplarity, the reproduction of whose texts serves as a nascent technology of domination for the consolidation of power within that state.

The argument can be taken one step further. Both the work of the humanist republic of letters and the growing missionary efforts of the post-Tridentine Catholic church indicate that a functional relation exists between the dissemination of universalizing ideology and the onset of European colonialism. In that case, the colonialist aspect of humanist texts like Milton's epic are the necessary obverse of their humanist affiliations—as well as of the changing register of humanism in the seventeenth century. In fact, *Paradise Lost* can be read retrospectively to mark the transmutation of "humanism" from the historically specific textual and educational practices characteristic of the early Renaissance to the abstractions whose project is the understanding of "man," in perfection as well as in disobedience, in "New World" as well as in old.[25] Adam and Eve, for instance, are bound to ideological double business: if the epithets "our author" (5.397) and "our general mother" (4.492) write human procreative history in a typical hand, then the end of the poem leaves these generic characters on the threshold of individuated experience, the career of the everyday, laboring fallen, whose "middling" ideological register is mapped out by the novel.

But as Foucault and others have suggested, "man" as the ineluctable Subject of History is precisely a historical phenomenon, a production of the classical episteme that emerges in post-Renaissance cultural formations.[26] If, as a consequence, the dissemination of "universal truths" falls within the register of incipient modernity, it stands to reason, first of all, that *Paradise Lost* enacts such a dissemination, and its theological plot carries along with it a notion of universal human nature capable of inflection by an alternative ideological formation. Second, however, it also suggests that, in a poem which commemorates destabilization at least as much as it does an overarching providential order, the authoritative production of universal history may be offset—indeed, broken into—by the unruliness, the discursive intransigence, of contemporary historical formations.

In that case the presence in *Paradise Lost* of Galileo and the optic glass suggests the pressure of an alternative modeling of universal truth, to which the poem can be cast as an almost dialectical response: the universals of a nascent scientific regularization. As chapter 5 argues, Galileo's *Dialogue on the Two Chief World Systems* must be read as an illustrative countertext to Milton's epic. The astronomical treatise yields a system of truths placed in a competing idealized space, the space of experiment in thought, fashioned to be free both of physical and ideological constraints. The universals generated within this utopian space are increasingly disembodied, increasingly "pure."[27] In contrast, the Miltonic version of ideality tends to instability, undone by femininity and the particularity of bodies, which are shown to be much the same thing.

Given the forbidden fruit, Catherine Belsey has suggested that misplaced desire in *Paradise Lost* is figured as oral.[28] But that assessment does not go far enough. Illicit longings, in paradise and out of it, work at many openings into the body, exploit the vulnerability of its many orifices. Consider besides the eating of the fruit, Satan "squat as a toad, close at the ear of Eve" (4.800); the eroticized dream he inspires; and the consuming ravages of Death. The quelling of right reason and faith by the demands of the body was a threat imaged in *The Tempest* by Caliban's disruptive, yet finally manageable, physicality. Here, it introduces into Milton's epic the possibility of grotesque subversion, a subversion both of genre and discourse. After all, it is the rebel angels who delve into the bowels of the firmament and who threaten to turn the war in heaven into a triumph of fecal evacuation in Book 6, a moment that in turn nearly destabilizes the poem's thematic seriousness.[29]

Perhaps the limit-case linking orifices to transgression is the terrifying absence of corporeal boundaries figured by Sin. Her openness to rape, incest, and base appetite marks her as troublingly permeable, and it suggests as well that for Milton it is a specifically female body that most registers the ravages of error. In contrast, consider the downward career of Satan: although he ends the poem a snake—after, of course, having passed through various other degenerative forms and resemblances—the poem has left open the possibility of (mis)reading him solely as the magnificent, ruined archangel at the beginning of the poem. Hence the purchase on the satanic of Blakean romanticism.[30] Whether the durable icon of Satan's splendid form is a matter of text or of a stubbornly modern inability to imagine the masculine body as changeable remains to be seen. However, as these summary instances make clear, in general the body under siege in *Paradise Lost* is always feminized—even when it is Adam's body. His excessive love for Eve causes him to fall, and to fall—as Samson was to do—potentially into an effeminating because uxorious desire.[31]

In drafting the script of a universal human history, *Paradise Lost* shows itself concerned to redeem the unlapsed body, the body "itself," from the charge of material debasement. Angels, for instance, have a species of body, which must be nourished like human bodies; as Raphael tells Adam:

> . . . food alike those pure
> Intelligential substances require,
> As doth your rational; and both contain
> Within them every lower faculty
> Of sense, whereby they hear, see, smell, touch, taste,
> Tasting concoct, digest, assimilate,
> And corporeal to incorporeal turn.
> For know, whatever was created, needs
> To be sustained and fed. (5.407–415)

The similitude between angelic bodies and their earthly correspondents extends to other pleasures. As Raphael's blushing response betokens, however, the analogy between angel and human begins to approach opposition, for the experiences of the flesh dissolve into airy congress, disarticulated from any exhausting corporeality:

Whatever pure thou in the body enjoy'st
(And pure thou wert created) we enjoy
In eminence, and obstacle find none
Of membrane, joint, or limb, exclusive bars:
Easier than air with air, if spirits embrace,
Total they mix, union of pure with pure
Desiring; nor restrained conveyance need
As flesh to mix with flesh, or soul with soul.
(8.622–629)

The enjambment created by the word "Desiring" betrays the comparison as it makes clear the inextinguishability of angelic longing; encumbered as they are with "restrained conveyance" of skin and bones, humans and their pleasures seem limited, materially different even if metaphysically analogous. Angelic sex, like the variable gender of spirits detailed in Book 2, is more a principle of isomorphism than a parallel somatic experience.

This programmatic reclamation of the flesh, already founded upon historical instability, cannot survive the Fall. More pertinently, it cannot even survive the pressure exerted by taxonomies of difference emergent in, for example, the practice of anatomy. Thomas Laqueur has provided a useful study of the role of anatomies, dissections, and increasingly scientific models of the human body on the formation of the modern sex-gender system during the sixteenth and seventeenth centuries. As Laqueur shows, the work of William Cowper, Kaspar Bartholin, and others begins, albeit incompletely, to register the "incommensurability" of male and female bodies as a new fact, as a visual and empirical discovery about nature and its essence.[32] Ovaries were distinguished from testicles; the cervical canal, which as late as 1543, in the illustrations for Andreas Vesalius's *De Humani Corporis Fabrica* resembled the inverted penis of Galenic theory, was given a name of its own. This shift in episteme consolidates modern hegemonies of gender, in which the sexed body becomes all-determining of difference; but, in effect, that body ratifies difference rather than discovers it and is itself subject to the exigencies of culture:

No discovery or group of discoveries dictated the rise of a two-sex model, for precisely the same reasons that the anatomical discoveries of the Renaissance did not unseat the one-sex model: the nature of sexual difference is not susceptible to empirical testing. It is logically independent of biological facts because already embedded in the language of science, at least when applied to any culturally resonant construal of sexual difference, is the language of gender. In other words, all but the most circumscribed statements about sex are, from their inception, burdened by the cultural work done by these propositions.[33]

For all of its contingency, the medicalized body, like the "space" of Galilean astronomy to be examined in chapter 5, proposes a form of universalism. It offers an idealist account of the material disposition of the body, perfect and representative specimens rather than messy particularity—a fiction of distinctness and order familiar to anyone who has ever dissected a frog in high school. But that suppression of particularity has a history. Rather than, like Laqueur, stress the "culturally and historically specific notions of what is ideal" (166) in the transcendent body of modern anatomy, I want to draw attention to the very emergence of scientific idealism itself in matters of the flesh. It resonates with—complicates—the universalizing task of humanist literature like *Paradise Lost,* as the defining image of Adam and Eve in Book 4 makes clear.

The shift from what Laqueur has called the one-sex model of the body (typical of pre-Renaissance theories of difference) to the two-sex model (the modern dimorphism that inflects "male" and "female") is visible in *Paradise Lost* as the difference between angelic and human incorporation, between spiritual and material instantiations of universal gender. Angelic—or demonic—gender is infinitely flexible: "For spirits when they please / Can either sex assume, or both" (1.423–424). Unlike their intelligential counterparts, Adam and Eve figure forth essentially different bodies, essentially different natures. The notorious line, "Not equal, as their sex not equal seemed" (4.296), inscribes gender hierarchy in outward somaticism, even if the watching Satan is not yet privy to the further secrets of difference that might lie under the skin. Idealized as moral agents, as the originary parents of humanity, Adam and Eve are also idealized embodiments of male and female.

These parallel inscriptions of difference collapse under increasing pressure in the time circumscribed by the plot of the epic. Witness Adam's vulnerability to the blandishments of Eve, which marks him not as essential in his maleness, but as permeable, unfixed, potentially effeminated. And the orifices of a feminized body, a body subject to outside influence and hence to the divagations of the historical, map onto the permeabilities of the exemplary humanist text that *Paradise Lost* aims to be. In signaling human vulnerability to need and desire, these openings into, interfaces of, the world become the thresholds of transgression and loss. And the location of a fall into particularity, and so an absence of corporeal boundary that comes to be securely installed in the feminine.

The particular language of my analysis, of moral loss seen as absence and signified by a dangerously unbounded female body, might suggest a quasi-Lacanian modeling of subjectivity as the not-so-*felix culpa* in the text. Such a reading would have much in common with the analysis of Cindy Sherman's photography by Laura Mulvey presented in chapter 1. Here, Milton's telescope, like the lens of a camera, becomes phallic, a masculinized compensatory device for a body riven by the Fall. The original sin that inheres in every being since Adam is tied (and cannot but be, as the initial descriptions of Adam and Eve make clear) to biological essence—made visible in the poem as the gendered, hierarchical world of the symbolic, a world belonging to science and other cultural practices. The astronomical espionage of Galileo thus renders him a prototypical analyst of the visual, readily assimilated to the question of perspective, authority, and the gaze, of who looks at whom and why.

Regina Schwartz picks up on the importance of the poem's many acts of looking in light of feminist film theory; her work, like Mulvey's, further helps to model an argument inflected by Lacanian models of subjectivity and also to show where such an argument fails.[34] The readiness with which Schwartz has placed Galileo within the framework of cinematographic beholding speaks, at the very least, to the pervasive sense of spatialization in the poem that I have noted elsewhere. To the extent that *Paradise Lost* appears to give back a glimpse of the (definitively lost) plenitude of Eden, it also approaches a filmic obsession with producing the absent signifier. But Schwartz's argument insists on the identity of sight and sexual *pleasure* within the prelapsarian plenum— and of the identity between the absences of film and absences within a printed text. Moreover, the pleasure it foregrounds is primarily congruent with post-Freudian models of libidinality and voyeurism, with a scopic regime whose objects and affects are already constituted as an immutable truth. As I have suggested with reference to Bacon's *New Atlantis,* the subject of science, as well as the object of pornography, is shaped by watching and being watched. "The Uses of Utopia" demonstrates how parenthesized data—the work of print—aid in the formation of a scientific observer. Stressing that signification and the making of a subject are materially contingent clarifies where Schwartz's essay elides the historical.

Can the visual system of *Paradise Lost* be read using the same models of spectatorship that film technology has demanded? This can be the case only if, as Fredric Jameson has suggested, the visual is *"essentially* pornographic."[35] But such a position denies instruments their historicity, their role in shaping and producing subjectivity effects. It is not enough to state that reading a poem, even when conducted in the postcinematic moment, is not the same as watching a movie, where one's gaze is contextualized and embodied by the selectivity of the camera lens. The objection, in its stronger form, insists that the lens of Galileo's optic glass is not the same technological apparatus as the camera's lens, nor does it produce the same effects. The viewing that Galileo's lens makes possible, especially at the moment of its development in the seventeenth century, is less dependent on an accomplished and even medicalized sense of the pleasure of looking at bodies subtending the latter. As Linda Williams has shown, the power-pleasure-anxiety nexus, which Schwartz's analysis assumes, begins to operate in the late nineteenth century, when camera technology can both freeze the body and frame (simulate?) its clinical truth.[36]

For the argument about the telescope as discursive prosthesis that I have sketched to yield to a Lacanian modeling of incorporation and lack, therefore, there must already be a connection between seeing, truth, and, ultimately, pleasure, a knowledge of the body that subjects it to the rigors of anatomical regulation and that gains pleasure as a function of such regulation.[37] There are not quite those bodies in *Paradise Lost,* just as there are not quite telescopes as such. Nevertheless, as Schwartz's reading implies, these discursive formations haunt the text as informants of powers to come, as my argument about science and the dimorphic body is intended to suggest.

What signifies as the proper history for (bounding?) a text obviously cannot be determined by the mere exclusion of postscripted critical practices. But Galileo and the uses to which his glass can be put force the question of anachronism. In not acknowledging that one historical consequence of the poem is a burgeoning division between fiction and fact, Schwartz's assimilation denies the rupture that announces itself to a putatively unsophisticated reader. In a different sense from the one that she has speculated upon, the quasi-prosthetized body of Galileo introduces a cultural motion into the schema of *Paradise Lost.* It is, however, one of contraries—a contrary notion of corporeality or, rather, of disincorporation, of a body only in theory. And it is paradoxically the anachronistic irruption of this historical figure into a narrative of universal origins which reveals that tale's imbrication in, subjection to, historical particularity.

The question of incorporatedness and transgression raised by *Paradise Lost* cannot be separated from the question of alterity with which this chapter began. In locating the most problematic bodies within a matrix of the diabolical and the feminine, the poem suggests the "natural" relationship between them as agents and figurations of the heterodox. In Milton's text they mark absolute changes of state, and through them the epistemic break that separates "Renaissance" from "Restoration," or, as Leonard Tennenhouse and Nancy Armstrong would have it, the difference between a subject defined by class and one defined instead by an emerging model of labor.[38] The poem itself, familiarly the last text of the Renaissance canon although written after the English civil war, straddles that epistemic break uneasily, as I have been concerned to suggest. This is perhaps most evident in *Paradise Lost*'s representation of fallen subjectivity. Its allusive structure urges that the Fall be understood through tropes of colonialism, that the specific drafting—conscripting—of Adam and Eve into a modern problematic of opposition and difference commences with the aboriginal loss of cultural innocence.

Such tropes present epistemological claims, as a passage after Adam and Eve have eaten the apple makes manifest. The knowledge they betray of having fallen can be glossed as an entry into epistemological dualism, which is to say into a different and distinctively modern historical formation. Their awareness of their nakedness is specifically a knowledge of their flesh, based upon a visual inspection of it. Although shame as a cultural construct is not to be mistaken for a supposedly objective knowledge of matter, the fact that such objectivity operates itself as a code places both formations on contiguous ground. In effect, shame is itself the moral signifier that the postlapsarian couple for the first time possess minds conscious of, and therefore potentially opposed to, their bodies.

Further, it is a shame coded—naturalized—through a strikingly ethnographic comparison. In coming to the version of Cartesian consciousness I have just detailed, Adam and Eve are explicitly compared with the inhabitants of the New World:

> Those leaves
> They gathered, broad as Amazonian targe,
> And with what skill they had, together sewed,
> To gird their waist, vain covering if to hide
> Their guilt and dreaded shame: O how unlike
> To that first naked glory. Such of late

> Columbus found the Americans so girt
> With feathered cincture, naked else and wild
> Among the trees on isles and woody shores.
> (9.1110–1118)

In one sense there is nothing novel about the passage, either in the reference to the recent history of European exploration or in the connection it makes between one set of innocents and another. The analogy between Eden and the Americas, and hence between Adam and Eve and the native inhabitants, was already a Renaissance commonplace, either by implication (America was a golden world without modern self-interest) or by explicit illustration.[39] Consider as an example of the latter de Bry's engraving for Harriot's *Briefe and True Report of the New Found Land of Virginia,* where a depiction of the last moment before Adam and Eve fall is inserted just before a protoethnographical sequence on North American native customs (figure 9). The Genesis couple are depicted as notably distracted; presumably, this loss of tranquility, proleptic of their sin, speaks also to the incursion made into North American culture by the observant and symbolizing eye of the illustrator. As the graphic counterpart of the optic tube, it makes representation the equivalent of aboriginal loss.[40]

The mere declaration that a connection exists between Adam and Eve and the Americans whom Columbus encountered is therefore not the most urgent claim that the passage makes on the attention. Instead, the colonizing of Adam and Eve asserts by concrete reference the interplay of the universal and the historical that the references to Galileo's optic tube already make a problem of. If, as I have argued, the telescope demands to be understood as an irruption of contemporary structures of feeling into a universalizing plot, this particular comparison offers a specific and complementary fix. In *Paradise Lost,* at least, the universal fallen subject of Scripture must paradoxically be an early modern subject; the beginning of History as such is consequently the first moment of colonialism.

Yet the passage also seems unsure of *what* history it brings into

9. Theodor de Bry, Untitled (Adam and Eve), from Thomas Harriot, *A Briefe and True Report of the New Found Land of Virginia* (by permission of the Folger Shakespeare Library).

play—or, rather, which textual tradition, which modeling of the human past. For what precedes this evocation of the fallen pair as the natural inhabitants of South America is an allusion to the banyan, a type of fig tree that flourishes in the "real" India. The slippage from India to "Indians" in the text repeats a real-life slippage in destination that inaugurates the forms and structures of early modern colonialism; it is as if the revised script of exploration were fetched in to indicate the isomorphism between Columbus's error and Adam and Eve's.

This problematic coalescence of prior fantasies of travel with the corrective evidence of the New World has already been hinted at in the phrase "Amazonian targe." It is unclear which Amazons the poem refers to, whether the recently encountered warriors of the river basin, or their textual prototypes, the fierce and female combatants of classical antiquity. Although the latter might seem to correspond more accurately to the poem's humanist referential system, the subsequent appearance of Columbus in simile form supplants—or at any rate renders insecure—that identification. Antecolonial suppositions merge with, give way to, colonial avatars. This lack of fixity in the allusions, in the signifying system of the historical, corresponds to the poem's uncertain allegiances to its moral script: despite its proclaimed intention to justify divinity, *Paradise Lost* cannot but be caught up and by the structures of novelty it, like Donne's *Ignatius His Conclave,* seeks to demonize.

The protocolonialist affiliations in *Paradise Lost* are most visible, although less textually specific, in the poem's earlier books. There, Satan's search for the "Space" that "may produce new Worlds" (1.650) becomes an ambiguous ratification of modernity despite the poem's investment in contradictory, and literary, universals. Satan seems the textual embodiment of the emblem of Icarus whose reconfiguration during this period, from impious transgressor to precursor of the liberal agent of modern power-knowledge formations, Carlo Ginzburg has traced.[41] The poem intends to condemn the devil's audacity as part of its project of justification, but it can also be said to glamorize that audacity, simply in the narrative importance which it affords to his unprecedented journey. Thus unshored from his designated locus after disrupting the fixed celestial hierarchy of heaven—and so the naturalized social stratification that heaven models—Satan prefigures the sovereign (because mobile) subject of scientific ideology whose conceptual adaptability suits him for the work of innovation.

However, Satan's career manifests one important difference from that of the emblematic Icarus, a difference that locates his flights early in *Paradise Lost* more specifically in the regime of colonialist enterprises. Having failed in his attempts to wrest material power over the heavens, he turns to unknown space as the frontier of conquest:

> . . . But first whom shall we send
> In search of this new world, whom shall we find
> Sufficient? who shall tempt with wand'ring feet
> The dark unbottom'd infinite Abyss
> And through the palpable obscure find out
> His uncouth way, or spread his aery flight
> Upborne with indefatigable wings
> Over the vast abrupt, ere he arrive
> The happy Isle. . . . (2.402–410)

This characterization of Satan-as-explorer has resonances throughout his journey out of hell and through Chaos; just before he reaches Sin and Death, for instance, his "solitary flight" (2.632) along the "coast" of the infernal deep is compared to mercantile journeys across the Ethiopian Sea (2.636–643). His intended work is not the acquisition of "spicy Drugs" (2.640), however, but the less material traffic in souls—a trajectory whose end point precisely connects with the presentation of Adam and Eve's inauguration as colonial subjects:

> . . . one for all
> Myself expose, with lonely steps to tread
> Th'unfounded deep, and through the void immense
> To search with wand'ring quest a place foretold
> Should be, and by concurring signs, ere now

Created vast and round, a place of bliss
In the Purlieus of Heav'n, and therein plac't
A race of upstart Creatures, to supply
Perhaps our vacant room, though more remov'd.
(2.827–835)

The projected conversion of Adam and Eve to the devil's party through a logic of deficit reduction connects the colonial Satan with the colonial Roman Catholic church: its missionary enterprises in the non-European world were explicitly read by the English writer Samuel Purchas in 1614 as attempts to redress the loss of allegiance to Catholicism in Europe:

For as much as the Papists do usually glory in the purchase of a new World unto their Religion . . . they have a new supply with much advantage in this Westerne World of America; and they make this their Indian conversion, one of the Markes of the truenesse and Catholicisme of their Church, which hath gained . . . an hundred times as much in the New World towards the West, South, and East, by new Converts, as it hath lost in the North Parts by Heretikes.[42]

Hence the pertinence of the Paradise of Fools (3.474–497). Yet despite the homology, Satan's quest cannot simply be reduced to a continuation of anti-Catholic polemics by other means. While *Areopagitica* provides evidence that Milton figures post-Tridentine Catholicism primarily as a censoring and oppressive institution, I have indicated that Galileo, an inevitably Catholic transgressor, occupies a complicated place in Milton's poem.[43] In effect, Satan's successful conversions in Eden imply that its fallen subjects are the subjects of modernity: the sect of Satan depends on the self-authorization of Cartesian formations, which can also include the isomorphic universalisms of Catholic ideology.

To the extent that Satan's colonial voyage is legitimated by the textual space it occupies, by the innovation on the hexameral tradition it represents, it figures as the structural counterpart of Adam's intellectual curiosity about celestial bodies. To draw on the poem's own language of transgression, it is the foretaste of the fall into knowledge. The devil's visionary quest, the necessary textual figuration of his conquest of Eden as "new World," is answered in the ostensibly contained discussion of the heavens that takes place in the garden in Book 8. "Ostensibly," because Adam's audacity in asking about the Copernican hypothesis is but incompletely discouraged by Raphael's rebuke: "Heav'n is for thee too high / To know what passes there; be lowly wise. . . . Dream not of other worlds" (8.172–173; 175). In fact, the orthodox rhetoric that intends to correct such curiosity by assigning it an inferior place in a hierarchy of wisdom is belied by Raphael himself. The angel, it seems, cannot avoid speculating about the issue that got Galileo into so much trouble:

> What if the Sun
> Be Centre to the World, and other Stars
> By his attractive virtue and their own
> Incited, dance about him various rounds?
> Thir wandring course now high, now low, then hid,
> Progressive, retrograde, or standing still,
> In six thou seest, and what if sev'nth to these
> The Planet Earth, so steadfast though she seem,
> Insensibly three different Motions move? (8.122–130)

It is crucial for my argument that Raphael is not seen to assert Copernicanism. In fact, his words manifest great care to deny his statements any positive content: "Not that I so affirm" (8.117). The reiterated formula "What if" that punctuates his excursus into celestial motion, like the "As-If" of the philosopher Hans Vaihinger, seems close indeed to the hypothetical and politic discourse characteristic of Galileo's embattled inquiry, as chapter 5 will examine.[44] Both signal volatile emergent formations and function as coded representations of interdicted or problematic epistemological propositions. But here the censor seems less the proximate force of Roman Catholic hegemony than a function of the epic's contradictory relation to those emergent formations in which it is so caught up. In effect, while circumscribing access, Raphael still gives Adam the discursive right to the telescope, to become the prosthetic observer whose gaze punctuates and contextualizes the poem and undoes its universal human script.

The "wandering" that in *Paradise Lost* has such an uncertain valence hardens into heresy, error outright, in the context of Galileo's speculations. In *The Writing of History,* Michel de Certeau notes that the project of representing the past depends on the silence of the social body under examination, which can be made articulate only through a willed suppression of the present space of writing.[45] This attempt to make the social body of Milton's text give up its secrets is, like the "Unspeakable desire to see, and know" (3.662) that consumes Satan, a model for a practice that finds its "purest"—although still incomplete—realization in the evacuated free space of Galileo's *Dialogue on the Two Chief World Systems.*

5

Galileo, "Literature," and the Generation

of Scientific Universals

This chapter's title is meant to suggest the emergence of the substantive counters that are symptomatic of modernity as commemorated in later disciplinary formations like "literature" and "science." The rhetoric here, and perhaps elsewhere in this study, may seem to propose a too drastic elision of the process that constructs those modern substantives in and as discourse. Alternatively, it may suggest a confusion of a multiply determined preformation with the differences produced and ratified once discourses become disciplines and acquire institutional valence. But the historiographic practice that has guided my analysis since chapter 1 invests in the past to comprehend the problematic of the present: hence the appropriateness of beginning with Cindy Sherman's "History Portraits." Rather than efface my investment in the cultural manufacture of opposition, then, I want to foreground it.

To combat the problematic and fetishized emphasis on "difference" within various constructions of postmodernity it is necessary to interrogate a specific mechanism of difference that emerges with the emergence of modernity. Such an interrogation yields a historical account of "otherness machines"—not, as in Sara Suleri's original phrase, the portal to the exotic that the West has made of the non-European subject, but here those discourses and institutions that will come to yield Enlightenment taxonomies of culture. As my frequent references to Michel de Certeau have made clear, this study is interested in proleptic manifestations of opposition and otherness, as specifically constituted through the parallel signs of the New World and the New Science. It is true that binary models have had a lengthy currency in European philosophy, a currency not solely determined by the preformations of the late Renaissance. Ian Maclean, for example, has shown their utility for concepts of feminine difference operating in legal and medical as well as philosophical systems whose histories extend back before the Scholastics; Thomas Laqueur, too, has provided a complementary account of the cultural stakes in the "clinical" truth of the body that emerges from pre-Renaissance formations.[1] Nevertheless, it is also the case that particular ways of making opposites are inscribed in particular practices, and thus have particular ideological motility. Hence the pertinence of quickened models of binarism in understanding the late Renaissance of the sixteenth and seventeenth centuries, which is to say, during a time when new cultural dominants are being constituted.

Among such new dominants is a newly conceptualized, instrumental relationship to "the natural," visible not only in the concomitant working over of newly colonized territories, but in the emergence of institutionalized science, which often has taken the body, in its presence or absence, as a point of departure. Earlier chapters have been concerned with examining the appearance of these new dominants in texts that, for complex cultural reasons, have fallen into the domain of the literary. Among these, perhaps Bacon's *New Atlantis* most problematizes the assignment of texts to disciplinary fields. Even so, its location in "literature" is comparatively stable, which is not the case for his more programmatic, more "scientific" works.

However, the argument of this book demands to be extended into a textual domain not wholly governed by the modern conventions of literature, yet specifically pertinent to the formation of those conventions in the late Renaissance. Given its function as a shadowy intertext for Milton's universalizing epic, Galileo's 1632 *Dialogo . . . sopra i due massimi sistemi del mondo, Tolemaico, e Copernicano (Dialogue on the Two Chief World Systems, Ptolemaic and Copernican)* suggests itself as an appropriate instance. The prosthetic telescope that haunts Milton's text occupies a central place here, and here, too, it is a sign of anxiety. In Galileo's text, however, it is a sign, unlike the "optic glass" of the repaired postlapsarian body, that registers its effects through its absence from the *Dialogue,* from a recognition that it can neither show nor be shown. As this chapter indicates, the telescope, even in its "natural" home, is not an unequivocal agent of modernity. Rather, as the concretion of the theologically volatile Copernican hypothesis, the instrument defines the ideological limits placed on scientific idealization in the seventeenth century and, hence, the limits of the prosthetic and instrumental male body.

Despite its intertextual symbiosis with *Paradise Lost,* its share in mapping out the opposition between universals of fact and fiction, Galileo's treatise moves the analysis away from the specificities of English culture in the seventeenth century.[2] By translating the argument to the Continent, however, the *Dialogue* reminds us that the transformation of the cultural field is a pan-European phenomenon, like the humanism that it displaces and replaces. Appropriately enough, the *Dialogue* can be considered from the standpoint of rhetorical demonstration appropriate to humanist textual production, as the studies produced by Maurice A. Finocchiaro and Jean Dietz Moss, among others, have shown.[3] But such demonstrations have largely preceded from the a priori that Galileo's writing *is* science. I want rather to suggest that it *makes* science: it produces "scientific truth" as a newly available cultural dominant by drawing on the conventions and growing ideological valences of the literary and by defining the literary as an imperfect mode for the formation of a scientific subject. For my purposes, Galileo's rhetoric, his frame of humanist reference, is no mere device that aids the astronomer in the demonstration of proofs. More crucially, the frame is an index, a way to chart the emergence of representational categories that come to belong to science proper—and a way to consider the importance of "literary" forms as the Other of scientific ones, within the installation and maintenance of this particular avatar of modernity.

In 1610, while still living in Padua, Galileo was beguiled by the recent explorations of the heavens made possible with his newly constructed telescope. To record his observations in some form, the astronomer sent a cryptic transmission to fellow scholars, among whom was Johannes Kepler: SMAJSMRMJLMEPOETALEVNJPVNENVGTTAVRJAS.[4] When his mystified correspondents asked that the message be decoded, Galileo subsequently rearranged the letters into sensible Latin: *altissimum planetam tergeminum observavi.*[5] As the decoding suggests, the astronomer had observed the planet Saturn and believed, on the basis of telescopic evidence, that it resembled a triple star. Although the transmission is not, properly speaking, a cryptograph, like a coded message the puzzle proclaims its significance at the same time that it renders itself opaque to casual inspection. By the account of the nineteenth-century historian Karl von Gebler, Galileo chose such a mode of communication because he wanted to outmaneuver his competitors: the jumbled message, therefore, proclaims a mixed desire, both to publicize his discovery, and to forestall recognition of it, to reveal by obscuring.[6] The observation of Saturn was recent, and Galileo had had scant time to confirm it through repetition. On the other hand, there was an equal reluctance to lose priority for the discovery should it turn out to be accurate—as, in the event, it was not: Galileo's rudimentary telescope was subject to optical flaws, which caused him to mistake the rings of Saturn for stellar multiplicity.

Both in positing the detection of a hidden rational order and in its self-reflexive anxiety about surveillance, the transmission maps out the ideological contours of early modern science. That all orders of the cosmos are manifestations of an occluded rational structure is characteristic of post-Renaissance epistemes, one constituted by the very expectation that such a structure exists to be revealed through inspection, as the analyses of Bacon and Milton have variously suggested. What is of interest here, however, is less the inaccuracy of the deciphered message—or, for that matter, the redoubling of circumspection, the betrayal that the watcher himself feels watched—than the nearly covert sign "poeta" visible in the disorderly mix of letters. I propose to read it from this vantage as the return of the historical repressed, if the word "poeta" can be taken as an iconic reference to the culture of late Renaissance humanism, whose educational practices and modes of representation constitute the hegemonic intellectual training of the Italian elite.[7] The unexpected sign of the poet marks a lost entrance, a way back into the cultural formation out of which modern scientific practice emerges. Anagrammatically, the letters position that emergent practice within a contemporary culture dominated by humanist notions of textual priority. This positioning has either been simplified because of nostalgia for an originary (and fictive) wholeness in culture, or else it has been suppressed by post facto constructions of disciplinary otherness—an alternative fiction of the scientific as always-already complete in itself.[8]

In part, this chapter complements the argument made in "The Prosthetic Milton" by suggesting that the emblematic and embedded trace of the poet in the Galilean code unconceals the differential production of modern ideology from within the hegemonic forms of late Renaissance humanism. In moving from the texts of a fully formed literary canon to a discursive field for which Galileo's *Dialogue* can be taken as exemplary, it seems pertinent to establish that this differential production of the scientific pervades Galileo's meditations on the practice of natural philosophy. Consider as a first instance Galileo's transmission to the German astronomer Johannes Kepler, dated 19 August 1610:

Verily, just as serpents close their ears, so do these men close their eyes to the light of truth. These are great matters; yet they do not occasion me any surprise. People of this sort think philosophy is a kind of book like the Aeneid or the Odyssey, and that the truth is to be sought, not in the universe, not in nature, but (I use their own words) by comparing texts. (Gebler, 26)

[Quid dices de primariis huius Gimnasii philosophis, qui, aspidis pertinacia repleti, nunquam, licet me ultro dedita opera millies offerente, nec Planetas, nec [sign for a moon], nec perspicillum, videre voluerunt? Verum ut ille aures, sic isti oculos, contra veritatis lucem obturarunt. Magna sunt haec, nullam tamen mihi inferunt admirationem. Putat enim hoc hominum genus, philosophiam esse librum quendam velut Eneida et Odissea; vera autem non in mundo aut in natura, sed in confrontatione textuum (utor illorum verbis), esse quaerenda. Cur tecum diu ridere non possum?] (Letter 379; X. 423)

In its ridicule of text-bound knowledge, Galileo's letter aligns the humanist practice of reading with the dominant Aristotelian school of astronomy, primarily advocated by the scholars who were, or would become, his opponents. While appropriate for the classical epic, the skills of philology and textual interpretation are, it seems, irrelevant, even counterproductive, in the investigation of natural phenomena. The rhetoric of "not in the world, not in nature" (*non in mundo aut in natura*) opposes world to text as a site of knowledge, and the self-convicting words of the reading philosophers stand against the protoempiricism of the community composed by Galileo, Kepler, and their admirers.

A similarly ridiculing passage occurs in the 1632 *Dialogue,* in connection with Simplicio's veneration of Aristotelian texts and the mode of study they demand: "he who hath thus studied him, knows how to gather from his Books the demonstrations of every knowable deduction, for that they contein [*sic*] all things."[9] In response, Sagredo burlesques this dedicated collation of textual fragments by comparing it to the practice of gathering choice passages of poetry for a commonplace book:

like as the things scattered here and there in *Aristotle* . . . that you perswade your self to be able by comparing and connecting several small sentences to extract thence the juice of some desired conclusion, so this . . . I could do by the verses of *Virgil,* or of *Ovid,* composing thereof *Centones,* and therewith explaining all the affairs of men, and secrets of Nature. But what talk I of *Virgil,* or any other Poet? I have a little Book much shorter than *Aristotle* and *Ovid,* in which are conteined all the Sciences, and with very little study, one may gather out of it a most perfect *Idea,* and this is the *Alphabet.* . . .[10]

It is clear that the methodological distinction which Galileo draws between world and text does not simply recast the comparison, offered by other seventeenth-century practitioners as well as by Galileo, between the Book of God and the Book of Nature.[11] Yet that comparison is a useful way to begin examining the discrediting of print-bound knowledge as it appears in emergent scientific ideology. The topos of the two divine books, for instance, is not merely metaphysical; rather, it can be read with specificity, imbricated as it is in the embryonic print culture of the Renaissance. As Chandra Mukerji has argued: "[The comparison] suggests that perceptions of nature and of natural processes . . . were linked to the development and use of printed books in scholarly research. When scholars not only used books in their work but also began to see their world *as* a book, they made this artifact a symbolic as well as practical artifact in European intellectual life."[12] Mukerji reminds us that books had but recently become widely disseminated objects, material contributors to the spread of knowledge throughout Europe. Further, her observation critically shifts the ground of textual significance away from purely scriptural exegesis to a recognition of a more general, and potentially "secular," tradition of print commentary. Accordingly, the texts to be read differentially in the Galileo passage are not alternative versions of the divine, each compelling its own form of ocular devotion. Rather, the operant conceptual distinction is between the methodologies of the *studia humanitatis* and those of the quadrivium, a wholesale partition between one type of textual practices and other, more "philosophical" ones, which would tend increasingly to dominate the former.[13]

For the purposes of this argument at least, it seems useful to complement such conceptual distinctions with potentially material ones. To invert Mukerji's rhetorical emphasis, there may well be an equivalent interest in concretion between humanist culture and that of natural science: each is interested in material objects, but those objects begin to assume unequal value and begin to circulate differently. The identification of classical astronomy with the collation of epics is thus reductive of the prestige of humanism. While working within a common trope of the text, the identification opens up a strategic difference between book-bound knowledge and the inspection of nature, with consequences not only for practice but for the creation of objects of study. By virtue of that difference, the "nature" subtended by emergent scientific practice becomes more real, more present to the inspecting eye, than the representational field of the classical text—and by implication the *exempla* of human nature that that text provides for study. It is not that the texts of Homer or Virgil themselves become disembodied: their "content," however, does. The division of intellectual work latent in Galileo's anti-Aristotelian polemic models a contrarian legitimation for scientific inquiry, where the Aeneid and the Odyssey, the foundations of a Renaissance *paideia,* have meaning primarily as negative occasions of the practice appropriate to natural philosophy.

Some other examples taken from Galileo's scientific writings, however, evince a rather less dismissive attitude to literary genres. In effect, these passages attest to the pervasiveness of humanist literary models as descriptive camouflage. Such references must be read against Galileo's laughing references to the classical epic in his letter to Kepler; in these other cases the operative distinction appears to be reading itself, which is to say the mode of practice for the apprehension of philosophical content. For instance, consider the specific language used in a letter of 23 May 1618, to Archduke Leopold of Austria that accompanied an earlier version of Fourth Day's material on the tides contained in the *Dialogue:*

With this I send a treatise on the causes of the tides, which I wrote rather more than two years ago at the suggestion of his Eminence Cardinal Orsini, at Rome, at the time when the theologians were thinking of prohibiting Copernicus's book and the doctrine enounced therein of the motion of the earth, which I then held to be true, until it pleased those gentlemen to prohibit the work, and to declare that opinion to be false and contrary to Scripture. Now, knowing as I do, that it behoves us to obey the decisions of the authorities, and to believe them, since they are guided by a higher insight than any to which my humble mind can of itself attain, I consider this treatise which I send you to be merely a poetical conceit, or a dream, and desire that your Highness may take it as such, inasmuch as it is based on the double motion of the earth, and, indeed, contains one of the arguments which I brought in confirmation of it. But even poets sometimes attach a value to one or another of their fantasies, and I likewise attach some value to this fancy of mine. . . . I have also let a few exalted personages have copies, in order that in case anyone not belonging to our Church should try to appropriate my curious fancy, . . . these personages, being above all suspicion, may be able to bear witness that it was I who first dreamed of this chimera.[14]

[Insieme con la presente, riceverà un mio breve discorso circa la cagione del flusso e reflusso del mare, il quale mi occorse fare poco più di due anni sono in Roma, commandato dall'Ill[ustrissi]mo e Rev[erendiss]mo Sig. Card[ina]le Orsino, mentre che tra quei signori teologi si andava pensando intorno alla prohibizzione del libro di Nicolò Copernico e della opinione della mobilità della terra, posta in detto libro e da me tenuta per vera in quel tempo, sin che piacque a quei Signori di sospendere il libro e dichiarare per falsa e repugnante alle Scritture Sacre detta opinione. Hora, perchè io so quanto convenga ubidire e credere alle determinazioni de i superiori, come quelli che sono scorti da più alte cognizzioni alle quali la bassezza del mio ingegno per sè stesso non arriva, reputo questa presente scrittura che gli mando, come quella che è fondate sopra la mobilità della terra overo che è uno degli argumenti fisici che io producevo in confermazione di essa mobilità, la reputo, dico, come una poesia overo un sogno, e per tale la riceva l'A. V. Tuttavia, perchè anco i poeti apprezzano tal volta alcuna delle loro fantasie, io parimente fo qualche stima di questa mia vanità . . . ne ho poi lasciate andare alcune copie in mano di altri Signori grandi; e questo, acciò che in ogni evento che altri forse, separato dalla nostra Chiesa, volesse attribuirsi questo mio capriccio, come di molte altre mie invenzioni mi è accaduto, possi restare la testimonianza di persone maggiori di ogni eccezzione, come io ero stato il primo a sognare questa chimera.] (Letter 1324; XII. 390.35–391.59)

The epistle proclaims the treatise "a poetical conceit" (*poesia*), a "chimera" to be "dreamed of" rather than an act of analysis and ratiocination. Perhaps more pertinently, the letter asks that the treatise—and therefore its analysis of the Copernican hypothesis—be received as such. Here, the mere claim of fictional intent is apparently protection enough from condemnation, just as the scrambled letters of his transmitted observations (which seem like a feeble protective device) were proof against intrusion. Indeed, under the cover of that intent, Galileo freely distributed the text to church officials, in part to safeguard it from claims of priority by Protestant astronomers. In part, however, the distribution also functions preemptively to insulate Galileo from claims of impiety and doctrinal recalcitrance: two years earlier, in

1616, the astronomer had been summoned to Rome and instructed by Cardinal Bellarmine not to "hold, teach, or defend" the Copernican hypothesis "in any way whatever, either orally or in writing."[15]

Much devolves on that injunction. Obviously, any later investigations published either formally or informally by the astronomer might be read as defenses of Copernicanism. But within the discursive conventions and contestations of seventeenth-century emergent scientific culture, a poetical conceit—which is what the letter avows the theory of the tides to be—could not be assessed for fidelity to doctrine.[16] Rather, the poetic here seems a register of alternative propositions, as though the text itself were the completing term of a metaphorical equation: merely to proclaim inquiry a fiction was enough, apparently, for it to be taken as such.

What more is at stake in the claim that a treatise on astronomy is a poetical conceit or a drama? More precisely, what type of interpellation does the text demand? It is obvious that, even from a seventeenth-century perspective, the two modes of literary production referred to are not interchangeable, at least not so far as the signification of genre is concerned. Yet their amalgamation suggests that the point is not generic but more generally discursive, necessitating a differential positioning of the reader. To identify astronomical propositions with poetical conceits seems to echo the contaminated textual fidelity of the Aristotelians, yet it turns that fidelity upside down. The difference is to be found where investigation ends and the politics of reception begins, since in the dedicatory epistle the emphasis shifts from the knowledge found in books as points of origin and authenticity, to strategies represented *by* books as forms of closure and foreclosure. To put it another way, it is only when scrutiny of nature ceases and the discourse of nature must circulate in the world that the poetical mode becomes appropriate.

But appropriate to what? Karl von Gebler feels such evasive stratagems reflect ill on both sides: "one really does not know which to be the most indignant at—the iron rule by which a privileged caste repressed the progress of science in the name of religion, or the servility of one of the greatest philosophers of all times in not scorning an unworthy subterfuge" (100–101). Rather than accept an analysis so imbued with positivistic assessments of the unquestioned value of "scientific truth," it is necessary to historicize the categories. Galileo's seventeenth-century translator, Thomas Salusbury, prefaced his English version of the *Dialogue* with a letter to the reader in which he is less harsh on the astronomer, preferring to blame the pope for his reaction to Simplicio and consequent persecution of Galileo: "thinking no other revenge sufficient, [Urban VIII] employed his Apostolical Authority, and deals with the Consistory to condemn [Galileo] and proscribe his Book as Heretical; prostituting the Censure of the Church to his private revenge" (*2r). Salusbury continues: "I shall not presume to Censure the Censure which the Church of Rome past upon this Doctrine and its Assertors. But, on the contrary, my Author having bin indefinite in his discourse, I shall forbear to exasperate, and attempt to reconcile such persons to this Hypothesis as devout esteem for Holy Scripture, and dutifull Respect to Canonical Injunctions hath made to stand off from this Opinion" (*2v). To a contemporary reader, Galileo's presentation appeared "indefinite"—not an evasion of a dangerous truth, nor the deceptions of a cowardly investigator. The very indeterminacy preferred in Salusbury's language is a reminder of the complex negotiations characteristic of an emergent formation—what Michel Serres's analysis, to be discussed at greater length in connection with the representation of Simplicio, has defined as the problematic of the third man.[17]

Since the *Dialogue,* on one level at least, has the advancement of a new cosmology and physics as one of its aims, it must attempt to set forth the arguments, proofs, and demonstrations upon which the Copernican system based its claim to legitimacy. Such an inquiry is both necessary and problematic, for reader as well as for author. The reader stands at one remove from it, witnessing not the proof itself but the representation of the proof (which is not always identical to its representation), convinced only to the extent that she can anticipate the responses and objections of Simplicio and Sagredo. The ocular evidence that the text provides can scarcely carry the same immediacy for the reader as it would if she could occupy the plane of the text. Indeed, the geometrical figures that compose part of the text underscore the distance—and difference—between it and its reader: the *Dialogue* demands that a diagram be accepted as the focus of a demonstration that accrues around it, never straying far from it. But the layout of the printed page and the temporal unfolding of the act of reading demand that the diagram be left behind to follow the argument. The figure becomes eccentric to the proof—a graphic representation that the space of reading is not the space of demonstration, that a scientific treatise cannot be addressed as one would a humanist narrative or dialogue.

Yet the allegiances to literary form persist and perform complicated work. During the Second Day of the *Dialogue* itself, Salviati, the most overtly pro-Copernican discussant, identifies the text with yet another literary genre:

And moreover we discourse for our pleasure, nor are we obliged to that strictnesse of one who *ex professo* treateth methodically of an argument, with an intent to publish the same. I will not consent that our Poem should be so confined to that unity, as not to leave us fields open for Epsody's, which every small connection should suffice to introduce; but with almost as much liber[t]y as if we were met to tell stories (142–143).

[Che anco, di più, discorriamo per nostro gusto, nè siamo obligati a quella strettezza che sarebbe uno che *ex professo* trattasse metodicamente una materia, con intenzione anco di publicarla. Non voglio che il nostro poema si astringa tanto a quella unità, che non ci lasci campo aperto per gli episodii, per i'introduzion de'quali dovrà bastarci ogni piccolo attacamento, e quasi che noi ci fussimo radunati a contar favole . . .] (VII. 188)

Stillman Drake translates the Italian "poema" as "epic" in his modern version of the *Dialogue,* thereby establishing, from the vantage of the twentieth century at least, the momentary equivalence—and hence discursive rivalry—between the Miltonic opus and the Galilean.[18] In the distance rhetorically permitted between the text and its near self-representation as "literature" can be located its most successful ventriloquism of that discourse. To put it bluntly, Galileo's *Dialogue* is a fiction because it pretends it is a fiction—but it is just that pretense (a nonpejorative word) which marks the emergence of a differential episteme.

From this perspective, it may be tempting to deem all such signification in Galileo a form of rhetorical (and hence conscious) camouflage, a strategy to express the interdicted, the marginal. Leo Strauss has made a similar case for key texts of Jewish and Islamic philosophy, whose authors, since their enterprise was deemed counterproductive of theology, were given to double writing—an overt reading for the masses, a covert one for the cognoscenti.[19] Indeed, the theoretical position Strauss articulates appears to be underwritten by the strong resemblance it bears to one Galileo himself advanced in his *Letter to the Grand Duchess Christina.* In responding to charges that the Copernican hypothesis was heretical because it directly contradicted biblical passages on the movement of the sun, the astronomer argued that Scripture was marked by a multiple sense of audience. While the Bible accorded with common observation in the instance of things indifferent—such as celestial motion—as the word of God it reserved for itself the right to be reinterpreted by church fathers in such things, as even in doctrinal issues: "If the Bible, accommodating itself to the capacity of the common people, has on one occasion expressed a proposition in words of different sense from the essence of that proposition, then why might it not have done the same, and for the same reason, whenever that thing happened to be spoken of?" ["Perchè, se occorrendo alla Scrittura, per accomodarsi alla capacità del vulgo, pronunziare una volta una proposizione con parole di sentimento diverso dalla essenza di essa proposizione, perchè non dovrà ella aver osservato l'istesso, per l'istesso rispetto, quante volte gli occorreva dir la medesima cosa?"; V.332].[20]

Despite the historical congruence and conceptual resemblance between the arguments, however, a similar hypothesis concerning Galileo's *Dialogue* seems inadequate both to any apparent intentions and to the complex means by which new cultural formations come into being.

For one, Galileo's use of vernacular Italian in the *Dialogue* is a gesture of inclusion, one that potentially extended its public circulation throughout Italy, unlike the Latin of the *res publica litterarum* meant for the community of European scholars. Second, to put faith in Strauss's hypothesis about hidden readings—even if the double text is produced by an astronomer given to anagrams—one would need to be convinced that the work of such references to the literary is entirely authorized, consciously produced by an oppositional political intent. The model for covert writing that Strauss employs cannot account for the complexity of an emergent discursive formation, even when located within the writings of a figure accorded exceptional status.

In choosing Italian for the *Dialogue,* Galileo in effect claims allegiance to a national vernacular literature rather than to a tradition of humanist scholarship, and thus to be following along the path of Dante, Ariosto, and Tasso. Hence the comparison of the conversation to an epic is, potentially at least, appropriate. At the same time, however, this ostensibly permeable, approachable text establishes itself at a remove from the literary signified by the epic, as something other than fiction; so the hypothetical subjunctive "as if" makes clear. What might be called the dialogue's citational mode becomes an index of polemical emergence, since to invoke the text as a fictional construct is virtually to inscribe it as something other, something more instrumental, than that construct.

In effect, the *Dialogue* deploys the genres of humanist writing as signifiers of its ideological innocuousness, as semaphoric disclaimers of its political agency. At the same time that the *Dialogue's* imbrication in humanist hegemony asserts an orthodox compliance with the rules of the game—which in this case are the injunctions neither "to hold, teach, or defend" Copernicanism—it is also constitutive of a new formation. The disciplinary boundaries that it marks between itself and the literary matter of humanism attest to the nascent division of cultural labor between "fact" and "fiction," "science" and "literature."

As I have suggested, modern taxonomies of writing that oppose "fact" to "fiction" themselves reside within the binary agenda of modernity—which, as Michel de Certeau has noted in *The Writing of History,* produces itself, makes itself "intelligible," through the conceptual mechanism of othering.[21] It follows that the opposition, whose rudimentary outlines can be glimpsed in Galileo's dismissive characterization of his opponents, can be read back into the *Dialogue on the Two Chief World Systems.* Within the history of science at least, the *Dialogue* has been canonized as a point of origin for the modern institution and its belief system; it might stand to reason, then, that the text would be at pains to assert its difference from "the Aeneid and the Odyssey." Yet time and again the *Dialogue* compares itself to precisely these (it would seem) discredited genres of writing—a fact that suggests a far more complex process of emergence for the oppositions I have mobilized in this discussion.

Consider the following passage from the Second Day:

Salviati: Before I proceed any farther, I must tell Sagredus, that in these our Disputations, I personate the *Copernican,* and imitate him, as if I were his *Zany;* but what hath been effected in my private thoughts by these arguments which I seem to alledg in his favour, I would not have you to judg by what I say, whil'st I am in the heat of acting my part in the Fable; but after I have laid by my disguise, for you may chance to find me different from what you see me upon the Stage. (113)

[Prima che proceder più oltre, devo dire al Sig. Sagredo che in questi nostri discorsi fo da Copernichista, e lo imito quasi sua maschera; ma quello che internamente abbiano in me operato le ragioni che par ch'io produca in suo favore, non voglio che voi lo guidichiate dal mio parlare mentre siamo nel fervor della rappresentazione della favola, ma dopo che avrò deposto l'abito, che forse mi troverete diverso da quello che mi vedete in scena.] (VII. 157–158)

Salviati impersonates a Copernican, plays his due part, rather than professes his strongly held beliefs. Given, however, that in effect Galileo's late friend is already playing a part, given that his name is circulating in the *Dialogue* as a sign for a certain type of inquiry, this reversion to the act and scene of scripted action underscores the importance of theatrical simulation.

It is possible to recognize contradictory moments in the *Dialogue* in which the bifurcations typical of the fully realized formations of modernity make their appearance—an opposition between discourses that predicate themselves upon true and false epistemological claims, like the scientific, and discourses which, like the literary, do not gain legitimation from such claims. Indeed, at times the text is explicit about its manner of constructing, and so defining, its preferred mathematical truths against a cultural repertory of alternative forms of signification. As Salviati's words during the First Day attest, the "Studies called Humanity" rely on the intervention of human ingenuity, of what can loosely be called "subjective" textual agency (and identified with the prone and vanquished forms of Aristotle and Demosthenes). In contrast, the philosophical task at hand needs but ordinary wit—and an ability to recognize the (objective?) difference between true and false:

If this of which we dispute, were some point of Law, or other part of the Studies called *Humanity,* wherein there is neither truth nor falshood, if we will give sufficient credit to the acutenesse of the wit, readinesse of answers, and the general practice of Writers, then he who most aboundeth in these, makes his reason more probable and plausible; but in Natural Sciences, the conclusions of which are true and necessary, and wherewith the judgment of men hath nothing to do, one is to be more cautious how he goeth about to maintain any thing that is false; for a man but of an ordinary wit, if it be his good fortune to be of the right side, may lay a thousand *Demosthenes* and a thousand *Aristotles* at his feet. Therefore reject those hopes and conceits, wherewith you flatter your self, that there can be any men so much more learned, read, and versed in Authors, than we, that in despite of nature, they should be able to make that become true, which is false. (40)

[Se questo di che si disputa fusse qualche punto di legge o di altri studi umani, ne i quali non è nè verità nè falsità, si potrebbe confidare assai nella sottigliezza dell'ingegno e nella prontezza del dire a nella maggior pratica ne gli scrittori, e sperare che quello che eccedesse in queste cose, fusse per far apparire a giudicar la ragion sua superiore; ma nelle scienze naturali, le conclusioni delle quali son vere e necessarie nè vi ha che far nulla l'arbitrio umano, bisogna guardarsi di non si porre alla difesa del falso, perchè mille Demosteni e mille Aristoteli resterebbero a piede contro ad ogni mediocre ingegno che abbia auto ventura di apprendersi al vero. Però, Sig. Simplicio, toglietevi pur giù dal pensiero e dalla speranza che voi avete, che possano esser uomini tanto più dotti, eruditi e versati ne i libri, che non siamo noi altri, che al dispetto della natura sieno per far divenir vero quello che è falso.] (VII. 78)

But these familiar bifurcations between "true" and "false" appear in that form, not because Galileo's text operates in a prior realm of fact and experience, but, paradoxically, precisely because it does not. The opposition is an ideological consequence of the *Dialogue*'s "literariness." The claim of literary intent (which might better be called the mimesis, even simulation, of the literary) turns the text's metaphorics of genre into a blind, a strategic frame for the politically and doctrinally volatile Copernican hypothesis. By repeatedly staging inquiry into the heavens as dramatic, epic, poetic, Galileo's *Dialogue on the Two Chief World Systems* produces literary modes as innocuous representation, in effect to screen its interdicted propositions about the movement of the earth. But the "truth" of science is not successfully occluded by poetry's "lies," nor is that the point of my analysis. Rather, a productive de-valuation of humanist forms is secured by the substitution of *fantasia* for fact that, as the papal reception of the *Dialogue* suggests, fooled no one for long.[22]

In a sense, conceptual binarisms have never been far from critical and historical discussion of Galileo's significance. The controversy occasioned by the publication of the *Dialogue on the Two Chief World Systems*, for instance, is frequently played as a drama of oppositions: the astronomer cannot but be positioned as its hero, since most interested scholars have invested in the typically modern authority of the scientific. Hence the historiography both replays and completes the canonical staging of the seventeenth-century trial. That canonical staging casts the astronomer in opposition to an ecclesiastical hierarchy both entrenched and benighted, whose dogmatic adherence to the letter of the Scripture led it to silence the forces of scientific truth and light.[23] Since the Copernican hypothesis had been condemned in 1616 as contradictory to the geocentrism depicted in the Bible, and since Galileo had been advised by Cardinal Robert Bellarmine not to profess the hypothesis, the publication of the *Dialogue Concerning the Two Chief World Systems* in 1632 indicated to the church Galileo's continued recalcitrance and heterodoxy. And the consequent repressive response of official church culture all but guaranteed that the astronomer would be deemed a martyr for scientific progress, one whose belated redemption from doctrinal limbo underscores the hegemony of the scientific in modern culture. Thus in 1979 Pope John Paul II, on the resonant occasion of the centenary of Albert Einstein's birth, called for the Galileo affair to be rethought in light of the "legitimate autonomy of science"; by 1992, Catholic doctrine exonerated the astronomer completely.[24] As a further gesture of compensation, the Vatican Observatory began to publish a series of monographs entitled *Studi Galileiani*.[25]

When the church accedes to the truth-claims of the scientific in the late twentieth century, it seems clear enough that the hegemony of that scientific must be approaching a new point of contestation. Yet even when the caricature of scientific modernity triumphing over doctrinal intransigence that I have just sketched is filled in with a more detailed hand, the resultant depictions have confined themselves to speaking of institutions more or less narrowly, as primarily religious or scientific. In fact, Pietro Redondi's attempt to shift the ground of Galileo's trial from astronomical hypotheses to an equally volatile atomism, and the skepticism that has greeted that attempt, speak eloquently to the canonical fixity of the historiography, at least as much as to any tenuousness in his argument.[26]

Yet despite the lingering triumphalism of many philosophical accounts, it has been equally possible to subject the astronomer to a charge of moral cowardice, for giving in to doctrinal coercion (and possibly to the threat of torture) and thereby betraying the absolutism of scientific truth at the moment of its discursive consolidation. That the unqualified status accorded to science is a post facto construction does not interfere with the either-or: in fact, it is precisely around such limit-cases that the ideology of disinterested, objective truth coheres.

Given that the positions he has been made to occupy are myriad, even contradictory, it seems clear that Galileo functions as a center of highly coded activity. Peter Armour, for instance, has written of "the accumulation of a myth [around the astronomer], universally meaningful through time and capable of constant examination and reinterpretation."[27] Similarly, Stillman Drake, perhaps the most significant contemporary Galileo scholar, characterizes the encounter between Catholic church and the astronomer as symbolic.[28] If we speak in terms of the historical mechanisms that produce discursive authority, however, Galileo becomes, not a mythic or symbolic production (which is to say a production outside history), but the equivalent of an author in the Foucauldian sense.[29]

It is certainly not for the continued utility of his writings to latter-day practitioners that Galileo warrants such designation. Foucault, for instance, explicitly debars Galileo qua physicist from access to the author-function: "A study of Galileo's works could alter our knowledge of the history, but not the science, of mechanics; whereas, a reexamination of the books of Freud or Marx can transform our understanding of psychoanalysis or Marxism."[30] The name of the astronomer is not, as Freud's or Marx's is, a signifier for the origins of a discursive practice, an origin that is, moreover, always "heterodox" and unassimilable to its later transformations. Hence the need in such cases for return to the author and to originary texts. While it is true that no one "returns" to Galileo as a way of rethinking motion, either local or celestial, perhaps his pertinence to ongoing laboratory and theoretical work in physics is not the point. Like the systematizing writers whom Foucault does designate as authorities, "Galileo" has proven anterior to a particular discursive practice, one that both seeks and creates the origins for the disciplinary boundaries of modernity.[31] So much is attested to by the Galilean allusions in *Paradise Lost.*

These references in Milton find their counterpart in Foucault's own reversion to Galileo, a reversion that in a sense completes the exclusionary project begun in the late Renaissance and documented by Milton's text. Given the historical problematic of which he is a part, it may be that Galileo is one of the few embodiments of scientific practice to whom the desire for mystique and prestige typical of the author-function could be attached, even when his suitability to the function is gainsaid.[32] This is simply to argue that an alternative canon to the one produced by ideologies of literature exists, and it consists of the great names of scientific history. This equivalence is, however, obscured by models of "revolution" or even of "normal science," which always presuppose the existence of "science" as an object of study.[33] Yet the prestige of membership in this alternative canon, like that of the more familiar literary one, is always in the process of reconsolidation, an activity that has little directly to do with the conduct of actually existing science. Even when "Galileo" signifies, as it usually does, within the unexceptional purview of the historiography of science, even when it is primarily an occasion for influence or source study, the name also functions as a critical location and the sign of an absence.

This is why Galileo constitutes an appropriate pendant to the analysis of Milton that precedes it. As authorized by the history of science, Galileo's trial, like Milton's epic, devolves on formal origin and ideological loss. Although the discipline of astronomy does not periodically reconstitute itself by a return to Galilean (or Copernican) foundations, his trial helps define the cultural field against which the future progress of scientific ideology is to unfold. The question, it seems to me, is less how to add to the extensive historiography devoted to the trial than how to step back from it, the better to consider what has itself been rendered exceptional, landmark, and thereby to locate the mythical, quasi-authorial status of Galileo in the preformations of the seventeenth century.

To accomplish this location, one must recognize that the texts of both Galileo and Bacon have a comparable liminality: their neither-here-nor-there status with respect to the world of letters separates them from direct (because literary) access to one definition of the author-function. Galileo, however, stands at one further remove from Bacon, the "Father of English Science," since the conjuring value of the former's name inheres in the importance of mathematics to some of his investigations as well as in the rhetorical utility of his writing for the instauration of scientific modernity. Galileo, then, uniquely represents both the equivalent (in seventeenth-century terms) and the Other (in terms of a modern division of cultural labor) of the canonical literary figures with whom this book has been concerned—particularly, it would seem, if Milton, his approximate counterpart. If Galileo and his glass percolate through Milton's major epic as alien signs of the history that its theological plot attempts to keep at bay, Galileo's "epic" (to use Drake's word) of universals reduces Milton's chosen humanist form to but one of a series of potential feints of signification.

For all of its significance in consolidating a nascent discourse around the investigation of nature, the *Dialogue,* unlike the other texts in this study, is not as frequently read as it is referred to. Often invoked as a cultural milestone, it has become perhaps as much a signifier, a site of origins, as the name of its author. This study is interested in the semiosis of the text as well as the text as semiotic event; for this reason, a brief introduction to and summary of the *Dialogue* are appropriate.

As the title indicates, this comparative exploration of the Ptolemaic and Copernican systems is framed as a conversation among three principles, Salviati, Sagredo, and Simplicio. In the address to the "judicious reader" ["al discreto lettore," VII. 29] that frames the text, Galileo identifies the first two as "Signor Giovan Francesco Sagredo, of a Noble Extraction and piercing Wit," and "Signor Filippo Salviati, whose least glory was the Eminence of his Blood, and Magnificence of his Estate, a sublime Wit that fed no more hungerly on any pleasure than on elevated speculation" ["Sig. Giovan Francesco Sagredo, illustrissimo di nascita, acutissimo d'ingegno . . . [ed] il Sig. Filippo Salviati, nel quale il minore splendore era la chiarezza del sangue e la magnificenza delle ricchezze; sublime intelletto, che di niuna delizia più avidamente si nutriva, che di specolazione esquisite"; VII. 30–31].[34] Both characters are named after friends who shared Galileo's interest in Copernican speculations. By the time of composition, however, Salviati and Sagredo had died; the text is therefore scripted, like Castiglione's courtly text of Urbino, to "perpetuate their lives to their honour." The elegiac here also meets the convenient: neither man could possibly be held to account for the utterances ascribed to him.

The text is subtly positioned in a time and space where cultivation follows upon leisure, which, as the comparison to Castiglione suggests, betokens a courtly milieu: in fact, the scene is identified as Sagredo's palace in Venice, a city in which Galileo had once lived. Access to this intellectual spaciousness—a term worth returning to—is, naturally, restricted, as the particularized affluence and high birth of Galileo's friends underscore. This particularity extends to characterological and analytical differentiation. Salviati, a Florentine who most often is taken to speak for Copernicus and hence for Galileo, is incisive, persuasive, and wry. Sagredo, although no less keen, represents the interested layman; he more often proposes objections or refines points of discussion.[35] But the third discussant, Simplicio, is not so clearly an embodiment of tributary mimesis, and this marks his character with a significant difference. With no single seventeenth-century referent, he is, apparently, a repository for whatever anti-Copernican and anti-Platonic stances the arguments necessitate at the time.[36] Occasionally, he reads or recites from Aristotle's *De Caelo,* his vade mecum, or from Scipio Chiaromonti's *De Tribus Novis Stellis,* an anti-Copernican treatise on comets and novae; at other times, he articulates the opinions of Galileo's ecclesiastical opponents. In any case, whatever he offers in the *Dialogue*—with the notable exception of the last day's disclaimer—is always dissected and disproved by his interlocutors, sometimes with exquisite politeness, sometimes with scorn.

The *Dialogue* occurs over four days, each day with its own topic and inflection. On the First Day the participants discuss the basic Aristotelian laws of nature that purportedly demonstrate the essential wrongness of Copernicus; by its end, Salviati has shown that there is neither a logical nor a physical necessity for Aristotle's restrictions on sublunary motion and the impossibility of superlunary imperfection. The Second Day is given over to an exploration of the geometrical nature of circular motion and to an analytic refutation of "commonsense" objections to the earth's movement. Why, for instance, doesn't a ball dropped from a high tower land many feet away from the tower rather than at its base, since the earth would have moved during the period of its fall? On the Third Day, Salviati undertakes to demonstrate that the heavens, contrary to Aristotelian dogma and to Chiaromonti, are indeed mutable, as astronomical observations of novae make clear. The brief discussion on the Fourth Day presents what was for Galileo an unassailable proof that the earth moves: the flux of the tides. He reasoned, albeit incorrectly, that as the earth moved it somehow dragged the oceans in its wake.[37] The text concludes with Simplicio's humbly expressed desire that the discussion of this one particular *"fantasia"*—a word inadequately translated by Salusbury's "conjecture"—not be taken as a limitation upon God's ability to bring about physical effects in whatever way he chooses, even if unsatisfactory to human reason.

This last day's discussion, along with the indeterminacy of Simplicio, betoken the ideological nexus within which the text is produced. Overtly, of course, he gives a habitation and a name—although not a historical one—to the institutionalized oppositions against Galileo, becoming their local representative. At the close of the dialogue he suggests that, however elegant and plausible the Copernican hypothesis has been shown to be, it may yet not be true, since elegance and plausibility are human values and may represent an imposition on God's omnipotence. The argument, since it emerges from the superior discursive authority of the theological, is unanswerable on its own terms, and so it closes the discussion. Simplicio's position, it has been demonstrated, represents Urban VIII's favored argument against the Copernican hypothesis, one whose inclusion was enjoined by the pope onto Galileo.[38] Nevertheless, it also represents a trouble spot within the text, as Giorgio di Santillana has recognized in terming it "Galileo's Folly."[39] On the one hand, Simplicio's bid for humility has all the authority of the last word; on the other, of course, that very authority is undermined by the characterization of the discussant who has uttered it. Simplicio has been ineffective throughout the dialogue, and his every attempt to uphold the Ptolemaic system—and by extension, the rule of Catholic hierarchy—has been depicted as illogical and inadequate. In effect, the *Dialogue* closes by making religion the last resort of the outmaneuvered.

One can join Santillana in wondering at Galileo's lack of adroitness, or alternatively defend the astronomer from the charge of dangerous obtuseness. Both of these positions, however, depend on reconstructing Galileo's intent, rather than on reading the *Dialogue* as a text riven with the ideological inconsistencies of the culture out of which it arises. Michel Serres's discussion of the "third man" in Platonic dialogue seems more apposite here.[40] In Serres's account of the dialogue form, the pure space of geometry depends on eliminating the incidental, the distracting particularity of the phenomenal: "I eliminate the empirical, I dematerialize reasoning. By doing this, I make a science possible, both for rigor and for truth, but also for the universal, for the *Universal in itself.* By doing this I eliminate that which hides form—cacography, interference, and noise—and I create the possibility of a science in the *Universal for us*" (69; emphasis Serres's). These stripped-down universals pertain most clearly to Galileo's own geometrical demonstrations, as the subsequent analysis shows. But the cacography to which Serres refers need not only be phenomenal, need not, that is, pertain solely to the contingencies and variabilities of matter. The noise may be the difficult contribution of the "third man," the commonsense empirical given figural advocacy: more than once, for instance, Simplicio protests that he is being asked to ignore the evidence of his senses in the device of geometry.[41] More broadly still, however, the cacographic third entity is the domain of the historical and ideological, of that which cannot be banished despite any impetus to the pure, which in fact defines that impetus. Thus Simplicio is a fictionalized embodiment of the white noise—or in this case the white writing—that constitutes the ideological conditions of the *Dialogue.* He represents, in the specific double sense of the word, the third man who is the proximate censor for the church; thus, he embodies the reproduction of conflictual elements from the text's post-Tridentine location. In a sense, the work of the *Dialogue,* the attempt to confound anti-Copernican arguments, is also the attempt to establish a utopian space for the conduct of early modern investigations of nature—a space, that is, free of prior inscription. Such a space must be purchased at some cost: hence the functional importance of leisure time and of the wealth that buys time for contemplation through patronage—or that underwrites the treatise's progression of analyses.

This analysis serves as prolegomenon to a reading of thought-experiments in Galileo's *Dialogue,* which I take as locations for the scripting of universals, the corollary, in some sense, of the literary universals of Renaissance to liberal humanism most pointedly examined in *Paradise Lost.* Humanist literary writing generates the ideology of "man" as a transcendent category. As Victoria Kahn has argued, the Renaissance practice of reading and reproducing classical texts positions the figures within those texts as exemplary instances of the human: consider Plutarch's *Parallel Lives* and the various uses made of Homeric or Virgilian figures.[42] The program that emerges from Renaissance philology takes that exemplarity as its ideological agenda in order to fashion the subject of humanist education, who can be said to inhabit a world in which all vestiges of historical material conditions—and hence of difference— are dispersed to the margins.

Galileo's textual experiments, embedded as they are in a dialogue that often asserts its near-identity to humanist literary genres, correspondingly give rise to, and naturalize, a complementary set of universals. If humanist textual modes interpellate the universal subject of history, Galileo's *Dialogue* bequeaths that subject such frame-independent concepts as "space." Indeed, by devising that space as precisely, disinterestedly "beyond" the pressures of seventeenth-century Catholic dogma, as I have suggested, the thought-experiments of the *Dialogue* make possible the canonical representation of science as an institution positioned outside the constraints of ideology, in a realm of pure thought analogous to the pure space of its most disembodied and de-materialized practices. Rather like Bacon's *New Atlantis,* it puts the problem of religious orthodoxy to one side, and so makes available another productive cultural division: between articles of faith and matters of fact.

The point of contact—not just between Bacon and Galileo but also between humanist pre-texts and scientific modeling—is the closeness of the thought-experiment to Renaissance utopian narratives. Through their effective isomorphism, the uninscribed space whose pretense of vacuity motivates the scripting of utopian texts becomes reinscribed within emergent scientific practice. Insofar as early modern science possesses a utopian dimension, this uninscribed, disinterested space at the fringes of early modern Catholic ideology is the space of universals modeled by the project of Enlightenment and by the liberal humanism that springs from its Renaissance avatar. But this instance of the utopian in early modern history is permeated by the contemporary forms of domination that it, like other utopian productions written in dialectic with early modern colonialism, seeks to banish from its purview. If its "interest" is not overtly political or authoritative, its effect, no less than Bacon's *New Atlantis,* is to interpellate the subject of early modern science in a space of universality—a universality with an oblique and potentially divisive relationship to the fictional, produced within a discursively colonized field.

Let me specify what I mean here in connection with staging the proof of the *Dialogue*'s arguments in favor of heliocentrism. Although the telescope is the sign of Galilean practice, it is conspicuously absent from the text except as a signifier of investigation that takes place "elsewhere." Somewhat like the geometrical diagrams with which the text is interspersed, the prosthetic apparatus is eccentric to the proof that the *Dialogue* lays out on its pages. Yet much depends on geometry and the telescope: they are the preeminent signs of a newly legitimated scientific practice.

The early modern optic glass, as Milton's *Paradise Lost* has amply demonstrated, is an unstable sign, a lack in technology as well as an aporia in representation. It has been argued by Paul Feyerabend that the early telescope was an undependable apparatus and no guarantor of proof; rather, it seems to have demanded the conceptual a priori of heliocentrism in order for it to underwrite any demonstration of heliocentrism's veracity.[43] Given the doctrinal strictures under which the *Dialogue* was produced, such conviction could not (be) materialize(d) through prosthetic means. The pro-Copernican Salviati could never have brought out a telescope, endowed himself with the augmented body of early modern science, because the anti-Copernican Simplicio could never have looked through one. This is not merely a problem in the rhetoric of proof, or of the layout of the page, both of which depend on the placement of the reader to be swayed by the arguments—although the material constraints of typography and the conventions of narration certainly signify here. Rather, the constraint is ideological. Simplicio could not be thus embodied, either as a believer or an unbeliever. If converted to belief, then Galileo had crossed the safe boundary of the hypothetical, of humanist literary pretext; if Simplicio were unconvinced after looking through the telescope, however, Galileo's text would have had to shut down, since the evidence of the telescope is a key element in demonstrating lunar imperfection, and hence in demonstrating that the heavens can be governed by the same physical principles as the earth. What Michel Serres has called the cacography of the world—the counterinscriptions of the historical and ideological—would have won out.

The *Dialogue* therefore moves to shut out the cacography, to produce a site of proof where "imperfections" are redefined, essentially, out of existence. Hence the location in the text of universalizing narratives, which have more generally, if anachronistically, been called "thought-experiments." The question of anachronism is important. Are such mental tests themselves a type of universal mathematics substitute, or are they as historically coded as other forms of textual practice?

According to Thomas Kuhn, the thought-experiment is a constant, an analytic tool that allows the scientist both to examine his own "conceptual apparatus" and to learn something new about the world he is attempting to conceptualize.[44] To illustrate this process of refinement, Kuhn chooses to analyze a demonstration from the First Day of the *Dialogue* in which Salviati attempts to show Sagredo and Simplicio the nature of velocity.

The demonstration is worth considering at some length in order to show where Kuhn's model itself eliminates the historical and reproduces the process it attempts to define. The demonstration in the *Dialogue* is prompted when Sagredo asks how a moving body can progress from a state of rest to a terminal speed without somehow getting mired in the process, since the "antecedent gradations of slowness" between the two states are infinite. Salviati answers that the moving body does not pause, even momentarily, at any one of these intermediate velocities. The concept of speed this response draws upon is subtle and counterintuitive, as the thought-experiment indicates. Salviati draws a right triangle, with base AB and hypotenuse AC, so that the perpendicular leg is CB. He then asks his two friends to consider what happens when balls are released along the inclined plane AC and the vertical distance CB: which ball moves faster? While the ball CB hits the ground first, there is no clear-cut answer, as it turns out, for it changes depending on what segment of the trip is being considered, whether near the initial point C, or at the final points A and B. Both balls arrive with the same final velocity, because both distances and times are proportional—as Salviati proves by drawing a perpendicular from B to line AC.

Kuhn stops his analysis there, rightly concluding that the thought-experiment has successfully allowed Galileo to demonstrate a more sophisticated idea of velocity, dependent on an abstract measure of time rather than on human perception. Galileo, however, brings the analysis back to the point of departure, planetary motion. Salviati asks the other two to consider motion along the incline AC as the height of CB approaches zero, that is, as AC comes more and more to approximate a horizontal surface. A ball dropped down an incline, be the incline ever so slight, will always roll before it comes to rest if no external forces intervene, whereas a ball placed along a flat surface, such as AB, has no tendency to move. Thus any steady horizontal motion has an angular component and is in reality circular motion around a center; it may once have had vertical and horizontal components, as with motion along a plane, but these components have since been converted to a uniform orbit.

This demonstration exactly fulfills Salviati's initial hypothesis. Through the "ruse" (the word is Serres's) of geometry, the superlunary is shown to map out onto the ordinary case of mundane matter:

We may upon good grounds therefore say, That Nature, to confer upon a moveable first constituted in rest a determinate velocity, useth to make it move according to a certain time and space with a right [i.e., straight] motion. This presupposed, let us imagine God to have created the Orb v.g. of Jupiter, on which he had determined to confer such a certain velocity, which it ought afterwards to retain perpetually uniform; we may with Plato say, that he gave it at the beginning a right and accelerate motion, and that it afterwards being arrived to that intended degree of velocity, he converted its right, into a circular motion, the velocity of which came afterwards naturally to be uniform. (12)

[Possiamo dunque ragionevolmente dire che la natura, per conferire in un mobile, prima costituito in quiete, una determinata velocità, si serva del farlo muover, per alcun tempo e per qualche spazio, di moto retto. Stante questo discorso, figuriamoci aver Iddio creato il corpo, v.g., di Giove, al quale abbia determinato di voler conferire una tal velocità, la quale egli poi debba conservar perpetuamente uniforme: potremo con Platone dire che gli desse di muoversi da principio di moto retto ed accelerato, e che poi, giunto a quel tal grado di velocità, convertisse il suo moto retto in circolare, del quale poi la velocità naturalmente convien esser uniforme.] (VII. 44–45)

Obviously, the line of argument is itself recursive; in pursuing it they have found themselves back at the point of departure, which is a discussion of the earth's potential motion. But the circling of the narrative also makes clear, defines, the space of inquiry. The three interlocutors begin their investigation by geometrizing the world of the material, reducing its forms to perfect instantiations and absolutes: as both Sagredo and Simplicio agree, a broken rock has its own form in perfection.[45] In the process of this alternative modeling of the ideal, they produce a space "beyond" (I use the term advisedly) the constraints of matter. Thus, imaginary balls run down imaginary inclined planes, much as they might appear to do in the real world, were conditions ever to approach the desiderata of mathematical demonstration. Initially, Salviati stipulates that the inclined plane and the ball to be imagined for this mental experiment be material objects, but so cleared of materiality that they approach the ideal:

if you had here a flat superficies as polite [sic] as a Looking-glass, and of a substance as hard as steel, and that it were not parallel to the Horizon, but somewhat inclining, and that upon it you did put a Ball perfectly Spherical, and of a substance grave and hard, as suppose of brass; what think you it would do being let go? . . . And how long would that Ball move, and with what velocity? But take notice that I instanced in a Ball exactly round, and a plain exquisitely polished, that all external and accidental impediments might be taken away. And so would I have you remove all obstructions caused by the Airs resistance to division, and all other casual obstacles, if any other there can be. (126–127)

[quando voi aveste una superficie piana, pulitissima come uno specchio e di materia dura come l'acciaio, e che fusse non parallela all'orizonte, ma alquanto inclinata, e che sopra di essa voi poneste una palla perfettamente sferica e di materia grave e durissima, come, v.g., di bronzo, lasciata in sua libertà che credete voi che ella facesse? . . . E quanto durerebbe a muoversi quella palla, e con che velocità? E avvertite che io ho nominata una palla perfettissimamente rotonda ed un piano esquisitamente pulito, per rimuover tutti gli impedimenti esterni ed accidentarii: e così voglio che voi astragghiate dall'impedimento dell'aria, mediante la sua resistenza all'essere aperta, e tutti gli altri ostacoli accidentarii, se altri ve ne potessero essere.] (VII. 171–172)

The conditions defined are avowedly perfect, as the diction makes clear. But it is not perfection in a Platonic sense, a sense, that is, that insists on an impossible realm of the ideational for its only success. Although thought-experiments are, practically speaking, as impossible to realize as any Platonic form, the emphasis they place is on perfection as a limit-case—a symbol—for the real. The elements of everyday, material life are posited only as hyperrefined, made capable of fulfilling the exigencies of an imagined geometrical demonstration used as an index of actually existing conditions:

Salv: Instance then wherein the fallacy of my argument consisteth, if as you say it is not concluding in the material spheres, but holdeth good in the immaterial and abstracted.

Simp: The material spheres are subject to many accidents, which the immaterial are free from. . . .

Salv: I very readily grant you all this that you have said; but it is very much beside our purpose: for whilst you go about to shew me that a material sphere toucheth not a material plane in one point alone, you make use of a sphere that is not a sphere, and of a plane that is not a plane; for that, according to what you say, either these things cannot be found in the world, or if they may be found, they are spoiled in applying them to work the effect. It had been therefore a less evil, for you to have granted the conclusion, but conditionally, to wit, that if there could be made of matter a sphere and a plane that were and could continue perfect, they would touch in one sole point, and then to have denied that any such could be made.

. . .

Salv: . . . I must tell you, that even in abstract an immaterial Sphere, that is, not a perfect Sphere, may touch an immaterial plane, that is, not a perfect plane, not in one point, but with part of its superficies, so that hitherto that which falleth out in concrete, doth in like manner hold true in abstract. And it would be a new thing that the computations and rates made in abstract numbers, should not afterwards answer to the Coines of Gold and Silver, and to the merchandizes in concrete. . . . Like as to make that the computations agree with the Sugars, the Silks, the Wools, it is necessary that the accomptant reckon his tares of chests, bags, and other such things: So when the *Geometricall Philosopher* would observe in concrete the effects demonstrated in abstract, he must defalke [*sic*] the impediments of the matter, and if he know how to do that, I do assure you, the thing shall jump no lesse exactly, than *Arithmetical* computations. (184–185)

[*Salv:* Assegnatemi dunque in che cosa consiste la fallacia del mio argomento, già che non conclude nelle sfere materiali, ma sì bene nelle immateriali e astratte.

Simp: Le sfere materiali son soggette a molti accidenti, a i quali son soggiacciono le immateriali. . . .

Salv: Oh tutte queste cose ve le concedo io facilmente, ma elle sono assai fuor di proposito; perchè mentre voi volete mostrarmi chc una sfera materiale non tocca un piano materiale in un punto, voi vi servite d'una sfera che non è sfera e d'un piano che non è piano, poichè, per vostro detto, o queste cose non si trovano al mondo, o se si trovano si guastano nell'applicarsi a far l'effetto. Era dunque manco male che voi concedeste la conclusione, ma condizionatamente, cioè che se si desse in materia una sfera e un piano che fussero e si conservassero perfetti, si toccherebber in un sol punto, e negaste poi ciò potersi dare.

. . .

Salv: . . . Ma io vi dico che anco in astratto una sfera immateriale, che non sia sfera perfetta, può toccare un piano immateriale, che non sia piano perfetto, non in un punto, ma con parte della sua superficie; talchè sin qui quello che accade in concreto, accade nell'istesso modo in astratto; e sarebbe ben nuova cosa che i computi e le ragioni fatte in numeri astratti, non rispondessero poi alle monete d'oro e d'argento e alle mercanzie in concreto. . . . Sì come a voler che i calcoli tornino sopra i zuccheri, le sete e le lane, bisogna che il computista faccia le sue tare di casse, invoglie ed altre bagaglie, così, quando il filosofo geometra vuol riconscere in concreto gli effetti dimostrati in astratto, bisogna che difalchi gli impedimenti della materia; che se ciò saprà fare, io vi assicuro che le cose si riscontreranno non meno aggiustatamente che i computi aritmetici.] (VII. 233–234)

To risk an awkward but important comparison, there is no nostalgia for an unfallen material world here, in contrast to Milton's many attempts to redeem matter from history and temporality, from the effects of aboriginal loss. The degree of confidence, even of satisfaction, in geometry manifested by Salviati, and hence by Galileo—and the importance of the telescope as a signifier of plausibility for the mathematical comparison—suggest the ultimate success of prosthetic novelty as a way to reconcile the heavens and the earth. Geometrical modeling becomes not merely a technology of knowledge, but a sign system with its own cultural logic, as the comparisons to coinage and bookkeeping indicate. Thus the Christian pastoral elegy is transformed into a ready acceptance of the world and its temporal, material relations. Instead of claiming to repair the defects of the Fall, as the Baconian program does, Galileo's geometry constitutes an alternative path to an alternative model of symbolic abstraction, a notion of representational efficacy that has its own universal purview.

The universal space that geometrical demonstration offers, however, has more in common with Baconian open-endedness and incompletion than might seem apparent at first glance. For all that geometry redeems matter (remember the perfect form of the broken rock), it cannot permanently guarantee its modeling. In the universalizing narrative of Galileo's thought-experiment all quotidian interference—cacography—is cleared out: balls glide unhampered by the hindrance of friction, by irregular conditions, by the imperfect conditions of the material, in effect if not in hypothetical demonstration. In parallel with the prosthetic eye of the telescope, Galilean science attempts to install the scientific subject in an artificial space of evacuated materiality, where indeed only an artificial body—a body without organs, a body only in theory—could last for long.

This space beyond, despite Kuhn's analysis, is historically specific. The unacknowledged contestation over the telescope (an instrument that is merely "theory materialized," in Bachelard's formulation, but that is really a signifier of ideological volatility) which shunts it from representation is one sign of what has been banished from the exemplary site of the *Dialogue.* For that matter, the homology I have already referred to between the scripting of this experimental space and the inscribing of the Americas within humanist utopias is another: the reiterated connection between New World and New Science in contemporary documents assures that colonialism and inquiry into nature be read as comparable tropes of domination whose work is the "clearing of space" typical of modernity's enterprises in novelty. Galileo's thought-experiments create a terrain free of ideological constraints, in which it is possible to discuss Copernicanism without fear of prosecution by the Catholic church. They stand in relation to the text of which they are a part as that very text stands in relation to seventeenth-century Florence: the conjunctive relationship is the fictive, the hypothesis of Copernicanism, the hypothesis of a free mental space which is produced as a possibility even as it travels under the sign of "as if."[46]

De Certeau discusses historiography, broadly conceived, as a written practice of meaning which replaces "the obscurity of the lived body with a 'will to know'" that must take textual form and that must, therefore, blank itself out in order to write. Thus he supplies a systematic model for the construction of modern systems: as he suggests, "'The making of history' is buttressed by a political power which creates a space proper . . . where a will can and must write (construct) a system," which he glosses as "a reason articulating practices" (6). But the practice of "blanking" that is the project of Galilean universalized narratives is not total. So much has been made clear by the last words of Simplicio, which constitute the *Dialogue's* so-called Fourth Day folly. At other moments, too, like the ones under consideration, the new cultural dominant of the scientific emerges: then the palimpsest of past scripts, the poet's trace in the empirical redrafting of the world, can be seen.

Let me take the argument to the concrete. Consider the following Second Day exchange among the three persons of the *Dialogue,* who are attempting to determine what would happen with a pen carried along by Sagredo on a sea voyage from Venice to Alexandretta (or Scanderon)—a common trade route for the Venetian empire: "If the neb of a writing pen . . . had had a facultie of leaving visible marks of its whole voyage, what signs, what marks, what lines would it have left?" (151). After the anti-Copernican Simplicio guesses that the pen would merely trace the path of the ship—and hence inscribe only a long, essentially straight line—he is corrected by Sagredo, who offers a more elaborate analysis of the problem:

If a Painter, then, at our launching from the Port, had begun to design upon a paper with that pen, and continued his work till he came to Scanderon, he would have been able to have taken by its motion a perfect draught of all those figures perfectly interwoven and shadowed on several sides with countreys, buildings, living creatures, and other things; albeit all the true, real, and essential motion traced out by the neb of that pen, would have been no other than a very long, but simple line: and as to the proper operation of the Painter, he would have delineated the same to an hair, if the ship had stood still. That therefore of the huge long motion of the pen there doth remain no other marks, than those tracks drawn upon the paper, the reason thereof is because the grand motion from Venice to Scanderon, was common to the paper, the pen, and all that which was in the ship: but the petty motions forwards and backwards, to the right, to the left, communicated by the fingers of the Painter unto the pen, and not to the paper, as being peculiar thereunto, might leave marks of it self upon the paper, which did not move with that motion. (152)

[Quando dunque un pittore nel partirsi dal porto avesse cominciato a disegnar sopra una carta con quella penna, e continuato il disegno sino in Alessandretta, avrebbe potuto cavar dal moto di quella un'intera storia di molte figure perfettamente dintornate e tratteggiate per mille e mille versi, con paesi, fabbriche, animali ed altre cose, se ben tutto il vero, reale ed essenzial movimento segnato dalla punta di quella penna non sarebbe stato altro che una ben lunga ma semplicissima linea; e quanto all'operazion propria del pittore, l'istesso a capello avrebbe delineato quando la nave fusse stata ferma. Che poi del moto lunghissimo della penna non resti altro vestigio che quei tratti segnati su la carta, la cagione ne è l'essere stato il gran moto da Venezia in Alessandretta comune della carta e della penna e di tutto quello che era in nave; ma i moti piccolini, innanzi e'n dietro, a destra ed a sinistra, communicati dalle dita del pittore alla penna e non al foglio, per esser proprii di quella, potettero lasciar di sè vestigio su la carta, che a tali movimenti restava immobile.] (VII. 198.17–199.4)

In claiming an imaginary space for inquiry into physical forces, this passage seems to foreground the autonomy of artistic production, as embodied in the form of the free-handed artist—who, for the purposes of allegorical simplicity, might stand in for the humanist author, the inscriber of "[una] storia."[47] As Sagredo assures the ever-naive Simplicio, whatever the path of the ship, the artist's pen represents the world independent of navigational progress. The humanist is inscribed in the text as a free agent, his activities unconstrained by the cosmic order that whirls around him. This poeta-surrogate ranges, not merely in the astrological zodiac of his own wit, as Sidney apotheosizes the producer of literature, but in an astrophysical system that appears to have no determining force over his inscriptions—inscriptions which range "per mille e mille versi" in thousands of directions or, perhaps, of verses.[48]

But then it becomes clear that the text subsumes such local motions of the hand—whether drafting trees and figures, or for that matter inscribing the narrative itself in the forum of literary forms—in a universal motion, produced in a space evacuated of all but what it calls the "true, real, and essential," a motion that would be the same even if the pen were merely carried rather than used. The shift in focus from agent to frame, from "art" to something that will travel under the banner of science, reduces the artist from autonomous to contingent, to a type of experimental cargo. The "real, true, and essential motion" of the frame is the "grand motion," "common" to "paper, pen, and all that was in the ship." By comparison, the pen-wielder's actions become "small," capable of leaving traces *precisely* because of their minuteness, because of their inability either to affect or be affected by the motion deemed grand.

In *Writing Matter,* Jonathan Goldberg has provided an account of the material formation of the subject of early modernity by and within "the hand" of writing, of literacy and discursivity.[49] Working with a post-Derridean model of the materiality of the signifier, Goldberg examines the standardization of penmanship, its disciplining and subjection of the body. Writing—that which writes the self—is both iterable and unique, a sign of uniformity and yet of individuality.

Although he does not cite them, Goldberg's analysis has much in common with the models of subject-formation regnant under modernity provided by Althusser and Foucault—specifically with respect to the way educational practices regulate the body while ostensibly training it in "expression."[50] Writing is not drawing, of course, although it could be argued that accomplishment in both is compatible with the humanist (and aristocratic) cultivation of sensibility. Yet what is common, and striking, in both Goldberg's account and the Galilean narrative is the sense that the free play of the line, of expression, is accomplished within an intricate yet naturalized system of constraints and forces. One system—that which enmeshes Goldberg's writing subject—is overtly ideological, overtly, that is, infused with the pedagogical agenda of Renaissance humanism. The other, of the casually drawing shipboard artist, is a system so naturalized—literally, so much the work of nature—that its operations can pass without remark.

It might be objected that to argue for the inconsequentiality of the artist's rendering is to mistake the larger point: that *all* local action—not merely that act of drawing, but for that matter the project of mercantile capitalism that informs the choice of itinerary—is rendered subsidiary, negligible in comparison with the universal motion of the earth. It may seem, too, that I am placing a great deal of weight on the rhetoric of scale, on the movement from small to large. Yet the allegory holds some explanatory utility. If the passage betrays the painter in the universalizing narrative, the "poeta" in the code, it also forecasts the increasing delegitimization of humanistic practice by proposing an alternative set of universals, of what is "real, true, and essential" in its place.

This sea voyage away from Venice has an incidental resonance, given Galileo's biography. Its historical counterpart in the astronomer's life is not a journey of mercantile profit bounded by the Mediterranean, nor the tropological exploration undertaken by a new Columbus; rather, a particular departure from Venice brought Galileo's conflict with Catholic hegemony into high relief. Galileo had lived in Padua, in the Venetian empire, until 1610, when the possibility of more lucrative patronage brought him back to Florence to become court astronomer to the duke of Tuscany. In returning to Florence from the Republic, which was comparatively independent of papal authority, Galileo placed himself in the path of submission. As Sagredo had remonstrated in 1611, rather prophetically: "Where will [you] find the same liberality as in the Venetian territory? . . . Who can promise with any confidence that, if not ruined, [you] may not be persecuted and disquieted on the surging billows of court life, by the raging storms of envy?" (Gebler, 32).

The choice to travel to Florence was certainly not a choice of a free space for inquiry, although the recompense may have offered some of the "liberality" that informs the otiosity and spaciousness of the *Dialogue*. The ship, if ship it was, that took Galileo away from the Republic also cut the path that led to the scripting of the *Dialogue*, and that path, as by now should be clear, could not pretend to be free from perturbation.

Conclusion: De Certeau and

Early Modern Cultural Studies

The preceding chapters are recognizably a contribution to a fairly traditional field, Renaissance literary studies. The readings that they offer of Shakespeare, Milton, and Bacon are motivated, at least in part, by strategies of close reading, a form of textual attention still privileged by English departments. I say "still," because the boundaries of disciplinary study are under siege at the present moment. Partly the siege has been launched, although not necessarily in a conscious or overt way, by U.S. universities themselves at the end of the twentieth century: a diminishing pool of resources forces traditional humanities disciplines to broaden their bases of operations, in effect to colonize other intellectual domains to justify their levels of institutional support.

Hence the expanded range of topics to be found in much literary scholarship, including the present work, must be read against the institutional horizons within which and in relation to which these studies have been conceptualized. But disciplines are also under siege for reasons proper to themselves, owing to ideological and epistemological critiques of intellectual business-as-usual that have accrued around the sign of "cultural studies." *New Science, New World* aims also to be read as a contribution to this emergent field. It offers a reading of emergent scientific culture that moves beyond the perimeters of "the body," which has recently become the favored site of engagement with the protoscientific in early modern studies. In fact, it is offered as a self-conscious, if incomplete, incursion into regions of "hard" science that, with few exceptions, have remained mystified entities to the humanities, even when they draw upon a discourse of constructionism such as that supplied by Foucault. To that end, the sections on Copernicus and

Galileo, in particular, are offered as examples of transdisciplinary work: I have aimed to do more than thematize science at the moment of its emergence, working instead to name and call into discourse some of the very structures and latencies—the historical a prioris—that make the emergence of a science of nature possible. My analyses are critical, therefore, in the way that history may be critical: to show how regimes of truth are consolidated, even those whose current epistemological agendas seem unimpeachable because their authority is so naturalized (as with mathematized sciences, for instance)—or, at any rate, so unreachable to scholars with humanities training.

Obviously, more needs to be done by those interested in the consolidation of modern dominants to interrogate the conditions under which the abstractions of mathematics came to power-knowledge: there are good political as well as scholarly reasons to breach that calcified division into "the Two Cultures." What, however, are the political stakes of early modern cultural studies? Given the dominance of mass culture— most people do not live "in the past" as readily as do scholars (and we generally visit only a tiny segment of it)—does historicist enterprise occlude or evade some of the engagements with forms of domination that gave urgency to cultural studies when it first arose? To what extent is it incumbent upon the study of early modernity to heed the debates around and critiques of knowledge-seeking agendas elsewhere, such as within the field of postcolonial studies?

I have not chosen my instance lightly, since the present study is being published during a period of heightened interest in early modern colonialism; this period of interest has stretched from approximately 1976, the American Bicentennial, past 1992, the notorious quincentennial of the Columbian voyages, and to the present. How may scholars understand their interest in the period of early modern transatlanticism apart from the legitimacy conferred by precedent—apart, that is, from the fact that a wealth of valuable work has been done and continues to be done on race and nationalism, mercantile and plantation structures, mapping, and other subjects, all under the aegis of early modern colonialism? Questions this broad are best responded to with local instances: thus the pertinence of Michel de Certeau to the present study, and indeed to early modern studies generally.

I begin, then, with the problem of an image, an act of representation. *The Writing of History* uses as its frontispiece the van der Straet engraving of the encounter between Vespucci and America that I discussed in chapter 2. Like much of de Bry, this engraving has become commonplace in studies of Renaissance colonialism. But little in the way of an overt reading of the image is offered by de Certeau, given over to a discussion of colonial narratives, nor to the material consequences of this representative inauguration of "discourse on the other." Indeed, the frontispiece finds its clearest echo in de Certeau's treatment of Jean de Léry: as nascent ethnography, Léry's "writing" becomes the privileged scene for the discursive capture of the non-European subject. At first glance, then, it seems that the emblematized scene of colonialism serves primarily as a figure, a trope, for concerns about the "larger" (i.e., more completely Old World?) formation of modern historical consciousness.

If to recognize the limits of de Certeau's study is to offer a critique of that study, the criticisms are familiar to readers of Gayatri Spivak's influential essay on the subaltern.[1] Spivak begins by reading a conversation between Foucault and Gilles Deleuze, in which they undertake to rethink the political function of intellectuals in late capitalism and to articulate the impossibility of reconstituting professional thinkers into a revolutionary vanguard for the proletariat.[2] But what Spivak hones in on is their concomitant assumption, based on the French thinkers' privileged access to modes of signification as the subjects of knowledge, that members of subordinated classes can speak for themselves and can be heard so to speak.

Although Foucault and Deleuze assert that "the masses no longer need [the intellectual] to gain knowledge" because "they *know* perfectly well . . . far better than he and they are certainly capable of expressing that knowledge" (207; emphasis in original), what is the basis for that certainty if not a recuperation of authority? Does "representation no longer exist" (206)—or have its difficulties been occluded and thereby multiplied precisely as a function of the self-consciousness of intellectuals about their power and privilege to represent?

The critique—and here the issues hinted at by de Certeau's frontispiece come into prominence—becomes the more pointed when the subordinated who are presumed able to express themselves are the subalterns of postcolonialism. To be sure, the inadequately attended-to voice of the subaltern cannot readily map out onto the textual subject of early colonial representation. Yet Spivak's general caution about the extension of the privilege to know and write into all domains, even as a function of critical advocacy, raises questions about the way practice, in working not to refetishize difference, might recapitulate the error of discursive dominance.

Any scholar who studies the formation of difference might well be chastened by the dilemma. It is surely the problem encapsulated by the van der Straet engraving and, indeed, the question raised within the present study, where colonialism is read tropologically, much as de Certeau does. It is possible, however, to read de Certeau's—and thus Foucault's—lacunae constitutively as the necessary, albeit unexamined, limits to the genealogical accounts they offer. In this dispensation, both Foucault and de Certeau stand as theoreticians of influential formations which it is critical to locate as specifically European, as in fact productive of "the West," understood itself as a regime of truth.[3] That this regime in its successive epistemological manifestations has constituted itself in relation to the silence of various social bodies—women, the mad, the mystic, the indigenous—cannot be denied. That these have been deemed Other seems to mean, in the writings of de Certeau, not that they remain opaque or remystified concepts, but that they were deemed subjugable within a system of knowledge whose historicity needs its own genealogy. By locating the quasi-genealogical project he undertakes in *The Writing of History* back in the sixteenth and seventeenth centuries—in the Renaissance that was both Foucault's own "lost chance" and, famously, the period whose scholars most often have engaged with Foucault in the U.S. academy—de Certeau begins to examine the conditions within which that system can be constructed.

Although his allegorized emergence of historical consciousness suggests that de Certeau duplicates the blind spots that Spivak has critiqued in Foucault, the redoubling serves a critical function. If the conditions under which a formation becomes visible are themselves subject to genealogical narration, then the formation has lost most of its productive force; it has ceased to signify in terms of the power-knowledge which operates alongside of nondiscursive formations. What succeeds the formation may be ideologically commensurable with its predecessor—may, for instance, as part of a discourse of "decolonization" be little different in repressive and material effects than imperialism in its prior avatars. But its new discursive power lies in its differential modality rather than in its survival in older, institutionalized forms. Survive it does, on local as well as global scales: despite the budgetary problems to which I initially alluded, traditional configurations of power-knowledge—whether in disciplines like "history" or "literature," or (still mystified?) periods like "the Renaissance"—exert specific centripetal pulls on scholarly research. These can accommodate themselves to geopolitical shifts or to shifts in the formation of global capital as such shifts figure distantly in the marketplace of ideas. Hence the popularity in Renaissance criticism of the van der Straet engraving, or, more broadly, the flurry of scholarly activity around early modern colonialism that accompanied the Columbian quincentennial.

De Certeau's own reading of Foucault, in his essay "Micro-Techniques and Panoptic Discourse: A Quid pro Quo," affords a clue for understanding the relationship of genealogy to the decline in productive power of its object.[4] He suggests that as the panoptic apparatus emerges from the "obscure stratum" that has heretofore been the unexamined space of its operation, "it might well find itself in the position of an institution itself imperceptibly colonized by other, still more silent procedures" (188–189). Its position has in effect become overt, "ideological": "As they have evolved, the apparatuses of surveillance have themselves become the object of elucidation and a part of the very language of our rationality. Is this not a sign that they have ceased to determine discursive institutions? They now belong to our ideology" (189).

They also belong to our history, as Foucault would of course recognize—or, rather, "our" history, with the inclusive possessive called to account. Not that history and ideology are self-identical. But in de Certeau's account, neither precisely engages with the operations of power directly: what is ideological is, like what is historical, based on available evidence. There are limits to historicist cultural studies as a form of advocacy in this regard: Foucault gave up writing the "archaeology of that silence" he undertook early in his career.[5] But if historicist work must be based on available evidence, like the ideology that structures rationality, it also determines what counts as that evidence in its very modality, its presumptions about a past quiescent enough, compliant enough, to comfortably be ventriloquized in discourse. Hence the value to early modern cultural studies of *The Writing of History:* like no other text I can think of, it forces us to pause in our scripting of the past, and to notice that we, too, are speaking power without knowing it.

Notes

Introduction

1. Michel de Certeau, *The Writing of History,* trans. Tom Conley (New York: Columbia University Press, 1988).

2. William Pietz, "The Problem of the Fetish, I, II, IIIa," a series of articles published in *Res* 9 (1985), 5–17; *Res* 23 (1987), 23–42; *Res* 16 (1988), 105–123.

3. Donna Haraway, *Primate Visions: Gender, Race, and Nature in the World of Modern Science* (New York: Routledge, 1989), 3.

4. See, for instance, "Situated Knowledges: The Science Question in Feminism and the Privilege of Partial Perspective" in *Simians, Cyborgs, and Women: The Reinvention of Nature* (New York: Routledge, 1991), 183–201.

5. Fernand Hallyn, *The Poetic Structure of the World: Copernicus and Kepler,* trans. Donald M. Leslie (New York: Zone Books, 1990). I regret that I discovered this book too late in my own researches to make adequate use of its extended analyses of two canonical figures in the history of science and of topics that in many ways parallel my own. His discussion of the hypothesis, for instance, has much in common with the abbreviated treatment I offer of the subject in chapter 1.

6. Max Horkheimer and Theodor W. Adorno, *Dialectic of Enlightenment,* trans. John Cumming (New York: Continuum, 1972).

7. Michel Foucault, Introduction to Georges Canguilhem, *The Normal and the Pathological,* trans. Carolyn R. Fawcett, with Robert S. Cohen (New York: Zone Books, 1991), 11–12.

8. *The Archaeology of Knowledge,* trans. A. M. Sheridan Smith (New York: Harper and Row, 1972), 57; see also 126–131.

1 Making It New

1. These are a series of more than thirty images made between 1988 and 1990, according to Thomas Kellein, "How difficult are portraits? How difficult are people!" in *Cindy Sherman 1991* (exhibition catalog; Basel: Kunsthalle, 1991), 10.

2. In a trenchant essay Abigail Solomon-Godeau has described how a growing consensus among mainstream critics about Sherman's importance as an image-maker has been accompanied by the erasure of gender as a category of analysis for her work. See "Suitable for Framing: The Critical Recasting of Cindy Sherman," *Parkett* 29 (1991): 112–115.

3. Laura Mulvey, "A Phantasmagoria of the Female Body: The Work of Cindy Sherman," *New Left Review* 188 (July–Aug. 1991): 147.

4. See Fredric Jameson, *Postmodernism, or, The Cultural Logic of Late Capitalism* (Durham, N.C.: Duke University Press, 1991); Jean Baudrillard, "The Ecstasy of Communication," in Hal Foster, ed., *The Anti-Aesthetic: Essays on Postmodern Culture* (Seattle: Bay Press, 1983), 126–134. For a critique of Jamesonian postmodernism that echoes mine in its attentiveness to feminist art practice and that deftly argues for its suspicion that such postmodernism conceals "a regret at the passing of the fantasy of the male self" (243), see Jacqueline Rose, "*The Man Who Mistook His Wife for a Hat* or *A Wife Is Like an Umbrella*—Fantasies of the Modern and Post-modern," in Andrew Ross, ed., *Universal Abandon? The Politics of Postmodernism* (Minneapolis: University of Minnesota Press, 1988), 237–250.

5. For example, Solomon-Godeau quotes Peter Schjeldahl's entranced response to the Untitled Film Stills: "I am responding to Sherman's knack, shared with many movie actresses, of projecting female vulnerability, thereby triggering (masculine) urges to ravish and/or protect" ("Suitable for Framing," 113).

6. Norman Bryson, "The Ideal and the Abject: Cindy Sherman's Historical Portraits," *Parkett* 29 (1991): 91–93. I have quoted the concluding lines of the essay.

7. Michel Foucault has provided the most sustained consideration on the relationship between clinical diagnosis and the penetration (if only conceptually) of the surface of the body. See *The Birth of the Clinic: An Archaeology of Medical Perception,* trans. A. M. Sheridan Smith (New York: Pantheon, 1973). Barbara Stafford's examination of the image repertory of eighteenth- and nineteenth-century models of the body works similar ground, but without the systematic considerations provided by Foucault. See Barbara Maria Stafford, *Body Criticism: Imaging the Unseen in Enlightenment Art and Medicine* (Cambridge, Mass.: MIT Press, 1992). Finally, for a study that, unlike Foucault's, emphasizes the work of gender in modeling the gaze of the scientific subject upon his object, see Ludmilla Jordanova, *Sexual Visions: Images of Gender in Science and Medicine Between the Eighteenth and the Twentieth Centuries* (Madison: University of Wisconsin Press, 1989).

8. This characterization of literary history is partly contradicted by New Historicism and cultural materialism. Indeed, each of these textual practices has sought to render traditional literary history problematic by taking subjugated

knowledges and forms into account, or by examining the mechanisms by means of which power (construed broadly through Foucauldian or Althusserian models) is consolidated and mobilized through early modern aesthetic texts. Even so, neither Stephen Greenblatt nor Jonathan Dollimore (to name but two representatives of these critical modes) is primarily interested in the Renaissance (pre)formation of the liberal-humanist model he writes against.

9. See Michel Foucault, "Two Lectures" in *Power/Knowledge: Selected Interviews and Other Writings, 1972–1977,* ed. Colin Gordon (New York: Pantheon, 1980), 78–108.

10. Thomas Harriot, *A Briefe and True Report of the New Found Land of Virginia* (1590; reprint with an intro. by Paul Hulton, New York: Dover, 1972), 75.

11. On the other hand, this may be the epidermal version of Elizabethan sumptuary laws—a displaced aristocratic body whose very extravagance of decoration confirms the racial superiority of the early British. I owe this point to discussions with Leonard Tennenhouse.

12. Richard Helgerson's *Forms of Nationhood: The Elizabethan Writing of England* (Chicago: University of Chicago Press, 1992) offers a more complicated account of the scripting of national consciousness in Renaissance England; Helgerson isolates a functional dialectic between antiquity and the Middle Ages as models for the narratives constituting the early modern state of England.

13. The genealogical connection between the illustrations is suggested (although not problematized) by Stephen Orgel in his edition of William Shakespeare, *The Tempest* (New York: Oxford University Press, 1987), 34–35.

14. I should note that, in general, sixteenth- and seventeenth-century discussions of the inhabitants of America speak of them as savages rather than, strictly speaking, as "primitives." Apprehension of the Other took place through a hypothesized privation from civilization, which had its benign form in fantasies of the "Golden World," where, as Pietro Martire insisted, "there is no mine and thine. . . ." For a further discussion of protoprimitivism in the Renaissance, see Hayden White, "The Forms of Wildness: Archaeology of an Idea," in *Tropics of Discourse: Essays in Cultural Criticism* (Baltimore: Johns Hopkins University Press, 1978), 150–182.

15. Johannes Fabian, *Time and the Other: How Anthropology Makes Its Object* (New York: Columbia University Press, 1983), 63.

16. I am alluding to Steven Mullaney's deft, influential, and rather problematic account of wonder-cabinets and, by extension, other productions of alterity and science in the English Renaissance. See *The Place of the Stage: License, Play, and Power in Renaissance England* (Chicago: University of Chicago Press, 1988), 60–87. For a trenchant critique of Mullaney's description of the wonder-cabinet as filled

with "things on holiday," see Amy Boesky, " 'Outlandish-Fruits': Commissioning Nature for the Museum of Man," *English Literary History* 58 (1991): 305–330.

17. Eugenio Garin, *Italian Humanism: Philosophy and Civic Life in the Renaissance,* trans. Peter Munz (New York: Harper and Row, 1965). Although Anthony Grafton in *Forgers and Critics: Creativity and Duplicity in Western Scholarship* (Princeton, N.J.: Princeton University Press, 1990) has argued that both forgery and scholarship give evidence that historical consciousness is a historical constant, he acknowledges that the apprehension of the past acquires a different inflection in the Renaissance. For a documentation of Garin's thesis, see Peter Burke, *The Renaissance Sense of the Past* (New York: St. Martin's Press, 1969).

18. Desiderius Erasmus, *Dialogus Ciceronianus,* ed. Pierre Mesnard; in Erasmus, *Opera Omnia* (Amsterdam: North Holland Publishing Co., 1971), esp. 1.2:624–630. See also Francis Bacon, *The Advancement of Learning, Book I,* in *Francis Bacon: A Selection of His Works,* ed. Sidney Warhaft (New York: Odyssey Press, 1965), 223.

19. *De rebus familiaribus,* XXI, 10, to Neri Morando, in *Letters from Petrarch,* trans. Morris Bishop (Bloomington: Indiana University Press, 1966), 170–171.

20. Michel Foucault, *The Order of Things: An Archaeology of the Human Sciences,* trans. of *Les Mots et Les Choses* (New York: Random House, 1970), 3–45.

21. See *Rome Reborn: The Vatican Library and Renaissance Culture* (exhibition catalog), ed. Anthony Grafton (Washington, D.C.: Library of Congress, in association with the Biblioteca Apostolica Vaticana [Rome], 1993), 112–115.

22. Grafton, *Forgers and Critics,* 26.

23. For a revealing analysis of forgeries from a sociological and curatorial standpoint, see Mark Jones, with Paul Craddock and Nicholas Barker, eds., *Fake? The Art of Deception* (London: British Museum, 1990).

24. Although the terms Renaissance humanism and liberal humanism have been used as effectively interchangeable by cultural materialist critics like Catherine Belsey in *The Subject of Tragedy: Identity and Difference in Renaissance Drama* (New York: Methuen, 1985) and Jonathan Dollimore in, for instance, *Radical Tragedy: Religion, Ideology and Power in the Drama of Shakespeare and His Contemporaries* (Chicago: University of Chicago Press, 1984), much historical work remains to be done examining the transition from one to the other. Lisa Jardine and Anthony Grafton in *From Humanism to the Humanities* (Cambridge, Mass.: Harvard University Press, 1986) offer some valuable hints about the shape of the argument in examining the institutionalization of humanism.

25. I have noted Jardine's and Grafton's account of Renaissance humanist training and its gradual transformation into a (purportedly) value-neutral technology for study in "the humanities." See *From Humanism to the Humanities.* For an introduction to the contents and culture of humanism, see Donald R. Kelley, *Renais-*

sance Humanism (Twayne's Studies in Intellectual and Cultural History 2) (Boston: Twayne/G. K. Hall, 1991); for a comprehensive anthology of contemporary scholarship on humanism, see Albert Rabil, ed., *Renaisance Humanism: Foundations, Forms, and Legacy* (Philadelphia: University of Pennsylvania Press, 1988).

26. Juliana Schiesari, "The Face of Domestication: Physiognomy, Gender Politics, and Humanism's Others," in Margo Hendricks and Patricia Parker, eds., *Women, "Race," and Writing in the Early Modern Period* (London: Routledge, 1994), 70.

27. *Petrarch's Letters to Familiar Authors*, trans. and ed. Mario Cosenza (Chicago: University of Chicago Press, 1910), 151 (from *De rebus familiaribus*, XXIV, 12).

28. See the preface to *The Great Instauration* in *Francis Bacon: A Selection of His Works*, ed. Sidney Warhaft (Indianapolis: Odyssey Press, 1965), 302–303. Unless otherwise indicated, all subsequent quotations from Bacon in this chapter are taken from this edition.

29. Stephanie Jed in *Chaste Thinking: The Rape of Lucretia and the Birth of Humanism* (Bloomington: Indiana University Press, 1989) speaks of the symbolic division between Lucretia's " 'violated body' and her 'innocent mind' " (13) as politically productive in separating philology from the political and the historical and hence in giving rise to the transcendental category "literature" as something apart from, and even eccentric to, the main political work of Western culture, which is always the reproduction of "liberty."

30. Grafton, *Forgers and Critics*, 26–28.

31. Warhaft, *Francis Bacon: A Selection*, 235–236. The classic, if oversimplified, account of Baconian gender is Carolyn Merchant's *The Death of Nature: Women, Ecology, and the Scientific Revolution* (San Francisco: Harper and Row, 1980). Since the publication of Merchant's book, the complex relationships between science and gender as historical discourses have been widely discussed. See, among many other sources: Nancy Tuana, ed., *Feminism and Science* (Bloomington: Indiana University Press, 1989); Sandra Harding, *Whose Science? Whose Knowledge?* (Ithaca, N.Y.: Cornell University Press, 1991); Page DuBois, "Subjected Bodies, Science, and the State: Francis Bacon, Torturer," in Michael Ryan and Avery Gordon, eds., *Body Politics: Disease, Desire, and the Family* (Boulder, Colo.: Westview Press, 1994), 175–191.

32. As chapter 2 on *The Tempest* will note, Miranda—who has actually brought Caliban to signification—is often elided in the narrative of this domestication. The elision calls to mind Petrarch's fantasy of an all-nurturing paternal classicism.

33. Samuel Purchas, from *Hakluytus Posthumus, or Purchas His Pilgrimes* 20 vols. (Glasgow: James MacLehose and Sons, 1905), I, 486; as quoted by Stephen Greenblatt, *Marvelous Possessions: The Wonder of the New World* (Chicago: University of Chicago Press, 1991), 10.

34. Purchas, *Hakluytus Posthumus,* I, 486, quoted in Greenblatt, *Marvelous Possessions,* 10.

35. This is not to claim, of course, that the inhabitants of the Americas lived in an eternal and naive present, but that the particular *form* of their historical archive—often pictorial, to be accompanied by a narrative that could change with the teller—was defined as inadequate and inaccurate. Walter D. Mignolo has written tellingly about the Western alphabet as the privileged vehicle for historical consciousness in early Spanish colonial narratives; as he suggests in connection with the little-respected Mexican codices, "According to the implied complicity [between alphabetical writing and history], it was concluded that anybody can keep records of the past, but history can only be written with letters" (219). See "Misunderstanding and Colonization: The Reconfiguration of Memory and Space," *South Atlantic Quarterly* 92, 2 (1993): 209–260.

36. John Donne, *Ignatius His Conclave* (Latin and English), ed. and annotated T. S. Healy, S.J. (Oxford: Clarendon Press, 1969), 5. Further citations to this text are supplied in parentheses.

37. Louis L. Martz has provided the groundwork for a bridge between Foucault's theoretical postulation and the specific forms of seventeenth-century English Protestant writing. See *The Poetry of Meditation: A Study in English Religious Literature of the Seventeenth Century,* 2d ed. (New Haven, Conn.: Yale University Press, 1962).

38. See "Mystic Speech," in *Heterologies: Discourse on the Other,* trans. Brian Massumi (Minneapolis: University of Minnesota Press, 1986), 80.

39. See Fernand Braudel's massive discussion of early modern world economies in *Civilization and Capitalism: 15th–18th Century,* trans. Sian Reynolds (New York: Harper and Row, 1981). Braudel makes a strong case for globalism in discussing the network of commercial and productive interactions typical of the fourteenth through eighteenth centuries; the highly articulated nature of the interconnections is, in Donne's text at least, offset by a perceived collapse of cultural categories between great things and small, between ideas and things. Compare Bacon's Idols, another attempt to theorize within a moralized structure (that of false worship) the cultural promiscuity of emergent capitalism. For case studies of particular novelties, see Chandra Mukerji, *From Graven Images: Patterns of Modern Materialism* (New York: Columbia University Press, 1983). For a broad-based survey of the problematics of consumerism in England in the seventeenth and eighteenth centuries, see John Brewer and Roy Porter, eds., *Consumption and the World of Goods* (New York: Routledge, 1993).

40. Marjorie Hope Nicolson, "The 'New Astronomy' and English Imagination," in Nicolson, ed., *Science and Imagination* (Ithaca, N.Y.: Cornell University Press, 1956), esp. 46–51.

41. See "High and Low: The Theme of Forbidden Knowledge in the Sixteenth and Seventeenth Centuries," *Past and Present* 73 (1976): 28–41. Ginzburg's examples do not concern colonialism, but rather what I have argued is its isomorphic counterpart in the natural philosophy of the seventeenth century.

42. Mullaney, *The Place of the Stage,* 61–62. I am unsure what is at stake for Mullaney in that suspended phrase "or suspension" in the passage I have cited; otherwise, his insistence on discontinuity is strategic, a way out, which I generally share, of a simple genetic model of cultural development. But insofar as the viewer of the wonder-cabinet has a postmodern body, Mullaney defines by absenting it the very relationship between the anthropologically estranged past and the familiar present.

43. Eilean Hooper-Greenhill, *Museums and the Shaping of Knowledge* (New York: Routledge, 1992), esp. 1–22; 78–132.

44. For examples of work on museum culture, see Ivan Karp and Steven D. Lavine, eds., *Exhibiting Cultures: The Poetics and Politics of Museum Display* (Washington, D.C.: Smithsonian Museum Press, 1991); Douglas Crimp, with photographs by Louise Lawler, *On the Museum's Ruins* (Cambridge, Mass.: MIT Press, 1993). Lawler has long been interested in the way museums operate to define art by negotiating the object for the viewer. For a 1989 exhibition at the Temporary Contemporary Museum in Los Angeles, she placed boxes of what appeared to be the explanatory cards conventional in other museums on the walls near other installations; these read, on one side, "It is something like," and on the reverse, "Putting Words in Your Mouth."

45. Margaret Hodgen, *Early Anthropology in the Sixteenth and Seventeenth Centuries* (Philadelphia: University of Pennsylvania Press, 1964), 111–161; see also Hooper-Greenhill, *Museums and the Shaping of Knowledge,* 30–39.

46. In the Ashmolean Museum in Oxford it is possible to see the relationship between the personal collection and the public museum. The museum started with the collection of Elias Ashmole, a seventeenth-century antiquarian; Ashmole's original collection has been reconstituted and set off in a complex coding of origins that is at once reprimitivizing (look at all this weird stuff jumbled together) and museologically imperial—the museum among whose exhibitions is the principle of its own origins. For a history of the collection, see R. F. Ovenell, *The Ashmolean Museum, 1683–1894* (Oxford: Clarendon Press, 1986); see also Boesky, " 'Outlandish-Fruits,' " 325–328.

47. Anthony Alan Shelton, "Cabinets of Transgression: Renaissance Collections and the Incorporation of the New World," in John Elsner and Roger Cardinal, eds., *The Cultures of Collecting* (Cambridge, Mass.: Harvard University Press, 1994); see 188–189 for a discussion of van Kessel.

48. Susan Stewart, *On Longing: Narratives of the Miniature, the Gigantic, the Souvenir, the Collection* (Baltimore: Johns Hopkins University Press, 1984), 136.

49. Certainly much of the research done on early modern collections suggests that scientific or natural objects come to predominate in number. See the essays collected by Oliver Impey and Arthur MacGregor in *The Origins of Museums: The Cabinet of Curiosities in Sixteenth- and Seventeenth-Century Europe* (Oxford: Clarendon Press, 1985).

50. Shelton, "Renaissance Collections," 198–203.

51. The authoritative account is R. F. Jones, *Ancients and Moderns* (1961; reprint, New York: Dover, 1982). For a corrective account of European humanism, see Anthony Grafton, *Defenders of the Text: The Traditions of Scholarship in an Age of Science, 1450–1800* (Cambridge, Mass.: Harvard University Press, 1991).

52. "To the Reader Concerning the Hypotheses in This Work," in *On the Revolution of the Heavenly Spheres,* trans. and intro. A. M. Duncan (New York: Barnes and Noble, 1976), 23.

53. Vol. II, Book III, General Scholium; the standard edition is Florian Cajori's reworking of Andrew Motte's 1729 English translation (Berkeley: University of California Press, 1962), 547. Newton's original Latin has occasioned much controversy. Motte—or perhaps Cajori—renders "fingo" as "frame," and the note to the line in Cajori's edition (n.55, 671–676) explains in great detail why Newton, who had obviously employed hypotheses throughout the *Principia,* would choose at the end of his text to repudiate them. More convincing is the argument advanced by Alexandre Koyré in "Concept and Experience in Newton's Scientific Thought," in his *Newtonian Studies* (Chicago: University of Chicago Press, 1965), 25–52. Koyré shifts the ground of explanation from the noun "hypotheses" to the verb *fingo, fingere,* which means "to feign" as well as "to frame." Newton found certain types of hypotheses suspect—for instance, that gravity was a physical property of objects— and these were the sort he disdained in the General Scholium. Koyré writes: "Thus, 'physical hypotheses' would be hypotheses in the very worst sense of the word, that is, false fictions which Newton quite properly refuses to *feign*" (39; emphasis in the original). It should be obvious that my reading of Newton is heavily dependent on Koyré's work.

54. Thus, Sextus Empiricus noted: "In one sense, [hypothesis] means the argument of a drama, as we say that there is a tragic or comic 'hypothesis,' and certain 'hypotheses' . . . of Sophocles and Euripides, meaning nothing else than the argument of the drama. And 'hypothesis' is used with another signification in rhetoric. . . . Moreover, in a third application we term the starting-point of proofs 'hypothesis,' it being the postulating of something for the purpose of proving something." Quoted by Wesley Trimpi, *Muses of One Mind: The Literary Analysis of*

Experience and Its Continuity (Princeton, N.J.: Princeton University Press, 1983), 26.

55. Trimpi, *Muses of One Mind,* 51.

56. Bernard Weinberg, *A History of Literary Criticism in the Italian Renaissance,* 2 vols. (Chicago: University of Chicago Press, 1961); Joel Spingarn, *Literary Criticism in the Renaissance* (New York: Columbia University Press, 1908).

57. Sir Philip Sidney, *A Defence of Poetry,* ed. J. A. Van Dorsten (Oxford: Oxford University Press, 1966), 23–24.

58. Richard Helgerson, *Self-Crowned Laureates: Spenser, Jonson, Milton, and the Literary System* (Berkeley: University of California Press, 1983). For a discussion of these issues as a problem of rhetoric, see the introduction by John Bender and David E. Wellbery to their edited collection, *The Ends of Rhetoric: History, Theory, Practice* (Stanford, Calif.: Stanford University Press, 1990).

59. For a thorough survey of the place of the hypothesis in contemporary astronomical discourse, see Ralph Blake, Curt J. Ducasse, and Edward Madden, "Theories of Hypothesis Among Renaissance Astronomers," in *Theories of Scientific Method: The Renaissance Through the Nineteenth Century,* ed. Madden (Seattle: University of Washington Press, 1960), 22–49.

60. For Luther, see Edward Rosen, *Copernicus and the Scientific Revolution* (Malabar, Fla.: Robert E. Krieger, 1984), 183; for Ramus, see Blake et al., *Theories of Scientific Method,* 43.

61. Blake et al., *Theories of Scientific Method,* 39.

62. Blake et al., *Theories of Scientific Method,* 43.

63. Quoted by Edward Rosen in the introduction to his translation of *Three Copernican Treatises* (New York: Dover, 1958), 24 n.68.

64. Robert S. Westman, "Proof, Poetics, and Patronage: Copernicus's Preface to *De Revolutionibus,*" in *Reappraisals of the Scientific Revolution,* ed. David C. Lindberg and Robert S. Westman (Cambridge: Cambridge University Press, 1990), 167–205. Westman is also interested in the patronage networks within which Copernicus, as a "clerical humanist" (175), would have circulated; this interest marks a significant intersection between the historiography of early modern science and the work of literary scholars such as Stephen Greenblatt and Louis Montrose. For another instance of the intersection, see Mario Biagioli, *Galileo, Courtier: The Practice of Science in an Age of Absolutism* (Chicago: University of Chicago Press, 1993).

65. *De Revolutionibus,* trans. A. M. Duncan, 22; Latin text in Copernicus, *Des Révolutions des Orbes Célestes,* trans. Alexandre Koyré (Paris: Blanchard, 1970), 27. All further citations to these editions are supplied parenthetically.

66. The phrase is Westman's. See "Proof, Poetics, and Patronage," 169–170.

67. Letter of Sir William Lower to Thomas Harriot, June 21, 1610, in Henry Stevens, *Thomas Harriot and His Associates* (London, 1900), 116–118.

68. *The Writing of History,* trans. Tom Conley (New York: Columbia University Press, 1988), 2–4; emphasis in original.

69. See A. D. Wright, *The Counter-Reformation: Catholic Europe and the Non-Christian World* (New York: St. Martin's Press, 1982), 121–146; and John O'Malley, "The Discovery of America and Reform Thought at the Papal Court in the Early Cinquecento," in *First Images of America: The Impact of the New World on the Old,* ed. Fredi Chiapelli (Berkeley: University of California Press, 1976), 1:185–200. For extensive studies of the Polyglot Printing Office, the Roman institution established to disseminate religious tracts in colonial languages, see J. Metzler, ed., *Sacrae Congregationis de Propaganda Fidei Memoria Rerum, Vol. 1: 1622–1700* (Rome: Herder Press, 1972).

70. Timothy J. Reiss, *The Discourse of Modernism* (Ithaca, N.Y.: Cornell University Press, 1982), 28–31.

2 Admiring Miranda and Enslaving Nature

1. Donna Haraway, *Primate Visions: Gender, Race, and Nature in the World of Modern Science* (New York: Routledge, 1989), 12.

2. Michel de Certeau, *The Writing of History* (New York: Columbia University Press, 1988), 2–3; Juliana Schiesari, "The Face of Domestication: Physiognomy, Gender Politics, and Humanism's Others," in Margo Hendricks and Patricia Parker, eds., *Women, "Race" and Writing in the Early Modern Period* (New York: Routledge, 1994), esp. 69–70.

3. Sander L. Gilman, "Black Bodies, White Bodies: Toward an Iconography of Female Sexuality in Late Nineteenth-Century Art, Medicine, and Literature," in *"Race," Writing, and Difference,* ed. Henry Louis Gates, Jr. (Chicago: University of Chicago Press, 1986), 223–261. Compare the poem "The Venus Hottentot" by Elizabeth Alexander (in *The Venus Hottentot* [Charlottesville: the University Press of Virginia, 1990]). Alexander's poem addresses, and redresses, the historical wrong done to Bartmann by imagining a voice and a history for her far beyond the merely illustrative and symptomatic uses to which the "Venus" is put by Gilman, in an essay that comes close to reinscribing the voyeurism it critiques.

4. The same point is effectively made about anatomy and physiology by Thomas Laqueur, *Making Sex: The Body and Gender from the Greeks to Freud* (Cambridge, Mass.: Harvard University Press, 1990); for a longer discussion, see chapter 4 below. The further point here, however, is that such critiques do not, for

Haraway at least, lead to a wholesale delegitimation of scientific inquiry. See her consideration of Bruno Latour's work in *Primate Visions*. As Haraway's other writing makes clear, devising a scientific practice responsive to critique and resistant to present ideological abuses is an urgent agenda for feminist scientists. See "Situated Knowledges: The Science Question in Feminism and the Privilege of Partial Perspective," in *Simians, Cyborgs, and Women: The Reinvention of Nature* (New York: Routledge, 1991), 183–201.

5. Haraway, *Primate Visions*, 133–185.

6. *The Tempest*, ed. Stephen Orgel (New York: Oxford University Press, 1987), 4.1.188–189. Further citations from this edition are provided in the text.

7. This is something of an engulfing topic. The historiography tends to be split between those who, like Frances Yates, accord great significance to the role of magic in Renaissance culture, and consequently to its influence on scientific practice, and those historians of "real," i.e., canonical, science, whose writings generally dispute the emphasis that Yates et al. place on hermetic and occult structures. For studies of Renaissance magic, see Frances A. Yates, *Giordano Bruno and the Hermetic Tradition* (Chicago: University of Chicago Press, 1964); Wayne P. Shumaker, *The Occult Sciences in the Renaissance* (Berkeley: University of California Press, 1975); Ioan P. Couliano, *Eros and Magic in the Renaissance,* trans. Margaret Cook (Chicago: University of Chicago Press, 1987). Brian Vickers usefully sums up the debate, and the pertinent historiographical evidence, in the introduction to his edited collection of essays, *Occult and Scientific Mentalities in the Renaissance* (Cambridge: Cambridge University Press, 1984), 11–55. However, the emergent formations of modernity are not necessarily as clearly defined as the latter-day taxonomic distinctions manifested in Vickers's introduction and in his choice of essays would suggest.

8. Robin Horton, "African Traditional Thought and Modern Science," as cited in Vickers, 33–44.

9. Horton, quoted in Vickers, 34.

10. A partial list would include the introduction to George E. Marcus and Michael M. J. Fischer, *Anthropology as Cultural Critique* (Chicago: University of Chicago Press, 1986), 1–6; Johannes Fabian, *Time and the Other: How Anthropology Makes Its Object* (New York: Columbia University Press, 1983); James Clifford, "Introduction: Partial Truths," in *Writing Culture: The Poetics and Politics of Ethnography,* ed. James Clifford and George E. Marcus (Berkeley: University of California Press, 1986).

11. Vickers, *Occult and Scientific Mentalities,* 9.

12. I use "unexpected" because it is clear that Vickers maintains an investment in the authority of the scientific. See "Analogy Versus Identity: The Rejection of Occult Symbolism, 1580–1680," in *Occult and Scientific Mentalities,* 95–163.

13. Michel Foucault, *The Order of Things,* trans. of *Les Mots et les Choses* (New York: Random House, 1970), 43–44.

14. See Marie-Rose Logan, "The Renaissance: Foucault's Lost Chance?" in *After Foucault: Humanistic Knowledge, Postmodern Challenges,* ed. Jonathan Arac (New Brunswick, N.J.: Rutgers University Press, 1988), 97–109.

15. Barbara A. Mowat has provided a valuable survey of these studies in "Prospero, Agrippa, and Hocus Pocus," *English Literary Renaissance* 11, 3 (1981): 281–303.

16. Eugenio Garin, *Science and Civic Life in the Italian Renaissance,* trans. Peter Munz (New York: Doubleday, 1969); Antonia McLean, *Humanism and the Rise of Science in Tudor England* (New York: Neale Watson Academic Publications, 1972); Pamela O. Long, "Humanism and Science," in *Renaissance Humanism: Foundations, Forms, and Legacy,* ed. Albert Rabil, Jr. (Philadelphia: University of Pennsylvania Press, 1988), 3: 486–512.

17. See Claude Lévi-Strauss, *Tristes Tropiques,* trans. John and Doreen Weightman (New York: Atheneum, 1981), 294–300; see also Jonathan Goldberg for a discussion of how Derrida reads Lévi-Strauss in *Writing Matter: From the Hands of the English Renaissance* (Stanford, Calif.: Stanford University Press, 1990), 3–6, 15–28.

18. See Edgar Wind, "Virtue Reconciled with Pleasure," in *Pagan Mysteries in the Renaissance,* 2d ed. (New York: Norton, 1968), 81–96.

19. Mowat, "Prospero," 287.

20. "Sexuality in the Reading of Shakespeare," in *Alternative Shakespeares,* ed. John Drakakis (New York: Methuen, 1985), 95–118, 95; further quotations appear parenthetically in the text.

21. A similar version of the essay, concentrating only on *Hamlet,* appears in Rose's *Sexuality in the Field of Vision* (London: Verso, 1986); see "*Hamlet*—the 'Mona Lisa' of Literature," 123–140.

22. A highly selective list of analyses would include Jan Kott's reading of the play as a recasting of the *Aeneid* informed by the imperialist subtext of classical Rome, "*The Aeneid* and *The Tempest,*" in *The Bottom Translation* (Evanston, Ill.: Northwestern University Press, 1987), 107–132; Stephen Greenblatt's positioning of Caliban as the victim of linguistic imperialism, "Learning to Curse: Aspects of Linguistic Colonialism," in *Learning to Curse: Essays in Early Modern Culture* (New York: Routledge, 1990), 16–39; and Terence Hawkes's casting of Caliban's possession of the dominant language as a political issue, "Swisser-Swatter: Making a Man of English Letters," in Drakakis, ed., *Alternative Shakespeares,* 26–46.

23. In this regard, critiques of the play from within Renaissance studies differ markedly from those emanating from postcolonial critics. While the first wave of

such analyses focused on Caliban, and to some extent fetishized him as official Other for the crop of Renaissance reading, postcolonial analyses that touch upon *The Tempest* have, after an initial emphasis on the slave, expanded their field to include Ariel, Miranda, and Sycorax. See, for instance, George Lamming, "A Monster, a Child, a Slave," in *The Pleasures of Exile* (1960; New York: Allison and Busby, 1984); Abena Busia, "Silencing Sycorax: On African Colonial Discourse and the Unvoiced Female," *Cultural Critique* 14 (1990): 81–104; and Ania Loomba, "Seizing the Book," in *Gender, Race, Renaissance Drama* (Delhi: Oxford University Press, 1989).

24. The groundbreaking essay in this regard is Lorie Jerrell Leininger, "The Miranda Trap," in *The Woman's Part: Feminist Criticism of Shakespeare,* ed. Carolyn Ruth Swift Lenz, Gayle Greene, and Carol Thomas Neely (Urbana: University of Illinois Press, 1980), 285–294; see also Laura E. Donaldson, "The Miranda Complex: Colonialism and the Question of Feminist Reading," in *Decolonizing Feminisms: Race, Gender, and Empire-Building* (Chapel Hill: University of North Carolina Press, 1992), 13–31; for Sycorax, see Stephen Orgel, "Prospero's Wife," *Representations* 8 (1984): 1–29.

25. Paul Brown, " 'This thing of darkness I acknowledge mine': *The Tempest* and the Discourse of Colonialism," in *Political Shakespeare: New Essays in Cultural Materialism,* ed. Jonathan Dollimore and Alan Sinfield (Manchester: Manchester University Press, 1985), 48–71.

26. George Lamming reads Ariel rather as "the embodiment—when and if made flesh—of the perfect and unspeakable secret police" (99), stressing the difference between lackey and slave, between being on the inside, as Ariel is, and being on the outside with Caliban. The reading is plausible, as well as poignant; still, it seems to attribute too ominous an effect to the spirit's work and to minimize the coerciveness of Prospero's dominion over it. See "A Monster, a Child, a Slave."

27. *The Tempest,* ed. Orgel, 69–70.

28. A version of this chapter, read at the Folger Shakespeare Library in the spring of 1988, occasioned much discomfort on the part of experienced scholars and editors in the audience. Although all agreed that assigning the line to Miranda had good textual authority, none thought the line could plausibly belong to her. Orgel's edition, however, does make a strong case for the "energy displayed" by the lines as "an important aspect of [Miranda's] nature" (120). Also see Loomba, *Gender, Race, Renaissance Drama,* 154–155.

29. Jan van der Straet, *Nova Reperta* (Antwerp, ca. 1600). The engraving, as my subsequent citations suggest, is more usually taken to be part of a series on America. But the boxed series in the Folger Library (Art vol. f81, 1600) has twenty plates numbered consecutively, and "America," numbered 1, follows the title page.

30. Given that the snake hoop resembles an ancient Egyptian symbol for eternity, the old man who carries it might be identified with Saturn or Father Time, both of whom carry that attribute. I am indebted to Sheila ffolliott for this information.

31. Louis Montrose, "The Work of Gender in the Discourse of Discovery," *Representations* 33 (1991): 2, 36 n.8. The edition cited by Montrose is a modern reprint: *"New Discoveries": The Sciences, Inventions and Discoveries of the Middle Ages and the Renaissance as Represented in 24 Engravings Issued in the Early 1580s by Stradanus* (Norwalk, Conn.: Burndy Library Publication no. 8, 1953). This set, also possessed by the Folger, contains modern glosses on the illustrations; in addition to "America," it provides several illustrations concerning the New World (nos. 1–4) not in the Folger's contemporary edition (which is also numbered consecutively). The Burndy Library reprint is a collation of two separate editions, the original run of twenty, an example of which is in the Folger, and a later four (title page and three others, of "Columbus, Vespuccius, and Magellan on their ships of discovery, shown in allegorical settings"), entitled *America Retectio* (Ben Dibner, "The New Discoveries of Stradanus," explanatory gatefold enclosed with the Burndy ed.).

32. This description, although drafted before Montrose's essay was published, parallels his in many ways; this note is thus an appropriate point to mention a more general indebtedness to Montrose's work on Elizabeth and the gendered body in early modern culture.

33. See Annette Kolodny, *The Lay of the Land* (Chapel Hill: University of North Carolina Press, 1975), esp. 10–14, for Renaissance fantasies of the feminized colony. More recently, Margareta Zamora has examined Columbus's inscription of "the Indies" within a domesticating discourse of the European feminine. See "Abreast of Columbus: Gender and Discovery," *Cultural Critique* 18 (1991): 127–149.

34. Montrose, "The Work of Gender," 4–5.

35. Joseph Hall, *Another World and Yet the Same: Bishop Joseph Hall's Mundus Alter et Idem,* trans. and ed. John Millar Wands (New Haven, Conn.: Yale University Press, 1981), 57–67. Although Wands's translation is modernized, he keeps the punning names of Hall's original text—unlike John Healey's 1609 translation into English, *The Discovery of a New World.*

36. There is no agreement about who is responsible for the striking engravings. For a review of the candidates (among them Titian and Vesalius himself), see *The Illustrations from the Works of Andreas Vesalius of Brussels,* ed. and annotated J. B. DeC. M. Saunders and Charles D. O'Malley (New York: Dover, 1973), 22–29. A useful and early recognition of Vesalius's importance can be found in Devon Hodges, *Renaissance Anatomies of Fiction* (Amherst: University of Massachusetts

Press, 1985), esp. 1–19. For the "theatrical" transaction between anatomist and corpse, see Luke Wilson, "William Harvey's *Prelectiones:* The Performance of the Body in the Renaissance Theatre of Anatomy," *Representations* 17 (1987): 62–95. A valuable discussion of the gendered opposition between knowledge-seeker and the object of his scrutiny is provided by Jonathan Sawday, "The Flaying of Marsyas: Dissecting the Renaissance Body," in *Renaissance Bodies: The Human Figure in English Culture, c. 1540–1660,* ed. Lucy Gent and Nigel Llewellyn (London: Reaktion Books, 1990), 111–135.

37. See Thomas Laqueur, *Making Sex: Body and Gender from the Greeks to Freud* (Cambridge, Mass.: Harvard University Press, 1990), esp. "New Science, One Flesh," 63–113. See also Nancy Tuana, "The Weaker Seed: The Sexist Bias of Reproductive Theory," in *Feminism and Science,* ed. Tuana (Bloomington: Indiana University Press, 1989), 145–171.

38. For an impressionistic but suggestive discussion of the ideologically motivated deployment of "women" as an analytical category, see Denise Riley, *"Am I That Name?" Feminism and the Category of "Women" in History* (Minneapolis: University of Minnesota Press, 1988).

39. Montrose, "The Work of Gender," 6–33.

3 *The New Atlantis* and the Uses of Utopia

1. See Michael McKeon, "Politics of Discourses and the Rise of the Aesthetic in Seventeenth-Century England," in *Politics of Discourse: The Literature and History of Seventeenth-Century England,* ed. Kevin Sharpe and Steven N. Zwicker (Berkeley: University of California Press, 1987), 35–51.

2. See the entry on "Literature" in Raymond Williams, *Keywords,* 2d ed. (New York: Oxford University Press, 1983), 183–188.

3. For a Foucauldian discussion of the epistemic shift from the Renaissance into scientific modernity as registered in utopian texts, see Timothy J. Reiss, *The Discourse of Modernism* (Ithaca, N.Y.: Cornell University Press, 1982).

4. Among many important studies, see Stephen Greenblatt, *Renaissance Self-Fashioning* (Chicago: University of Chicago Press, 1980); Louis Montrose, "Celebration and Insinuation: Sir Philip Sidney and the Motives of Elizabethan Courtship," *Renaissance Drama* n.s. 8 (1977): 3–35; Jeffrey Knapp, "Eros as a Means of Empire in *The Faerie Queene* I," *ELH* 54 (1987): 801–834; Maureen Quilligan, "Sidney and His Queen," 171–196; and Clark Hulse, "Spenser, Bacon, and the Myth of Power," 315–346, in *The Historical Renaissance,* ed. Heather Dubrow and Richard Strier (Chicago: University of Chicago Press, 1988).

5. From "Certain Considerations Touching the Plantation in Ireland. Presented to His Majesty" (1606), in *The Works of Francis Bacon,* ed. James Spedding, Robert Leslie Ellis, Douglas Denon Heath (New York: Garrett Press, 1968), 4: 123. All references to Baconian texts in this chapter are cited in parentheses.

6. See "Of Islands and Trenches: Naturalization and the Production of Utopian Discourse," *Diacritics* 7 (1977): 2–21, for Fredric Jameson's useful distillation of Louis Marin, *Utopics: Spatial Play,* trans. Robert Vollrath (Atlantic Highlands, N.J.: Humanities Press, 1984).

7. The argument that Bacon's texts generate scientific subjects is clearly much indebted to Stanley Fish's nuanced analysis of Bacon's rhetorical strategies in the *Essays.* See Fish, "Georgics of the Mind," in *Self-Consuming Artifacts* (Berkeley: University of California Press, 1972). But where Fish sees the transaction between reader and text occurring in an evacuated site of reading, inevitable because reiterable, I am concerned to fix the possibility of a historically contingent operation of the text. Thus *The New Atlantis* is less a rhetorical mechanism in perfect control of what it processes than a product and producer of a necessarily incomplete discursive innovation; as a result, contradictions and hermeneutic difficulties do not so much generate an effect of scientific skepticism as provide the traces of its emergence in the late English Renaissance.

8. The literature critiquing science as discourse or ideology has become extensive. See Max Horkheimer and Theodor W. Adorno, *Dialectic of Enlightenment,* trans. John Cumming (New York: Continuum, 1972). Caroline Merchant's *The Death of Nature: Women, Ecology, and the Scientific Revolution* (San Francisco: Harper and Row, 1980), is important, although its nostalgia for an immediate experience of the natural world, as well as its sense that a grassroots feminism can renew that enchanted world, derive from an essentialism now difficult to support. See also Paul Feyerabend, *Against Method* (London: Verso, 1972); Ilya Prigogine and Isabelle Stengers, *Order Out of Chaos: Man's Dialogue with Nature* (New York: Harper and Row, 1984); Nancy Tuana, ed., *Feminism and Science* (Bloomington: Indiana University Press, 1989); and Sandra Harding, "Feminism, Science, and the Anti-Enlightenment Critiques," in *Feminism/Postmodernism,* ed. Linda Nicholson (New York: Routledge, 1990), 83–106.

9. As the *OED* indicates, during the early to mid-seventeenth century, when modern science is in the process of emergence, the two words are used interchangeably; when science comes to constitute itself professionally—and when its practice is alienated from ordinary life—a sustained distinction between the two words appears.

10. Literature and science as categorical discourses (but not as the simplistic bifurcation of the "Two Cultures") are the latest products of a process of cultural

differentiation necessarily complex and difficult to summarize. Richard Helgerson and John Guillory have argued that the issue of authority is important to the Renaissance emergence of literature as a self-constituted discourse. This emergence, and its concern with cultural and representational control, bears upon the discursive maneuvers I am concerned to excavate within natural philosophy. See Helgerson, *Self-Crowned Laureates: Spenser, Jonson, Milton and the Literary System* (Berkeley: University of California Press, 1983), and Guillory, *Poetic Authority: Spenser, Milton, and Literary History* (New York: Columbia University Press, 1983).

11. Margaret Hodgen, *Early Anthropology in the Sixteenth and Seventeenth Centuries* (Philadelphia: University of Pennsylvania Press, 1964), esp. "The Ark of Noah and the Problem of Cultural Diversity," 207–253.

12. Jean-François Lyotard's *The Postmodern Condition: A Report on Knowledge,* trans. Geoff Bennington and Brian Massumi (Minneapolis: University of Minnesota Press, 1984), for instance, shows just how ineluctable the recourse to scientific precedence has become. Lyotard invokes the micronarratives of chaos theory to resolve the "legitimation crisis" in contemporary culture brought on by the obsolescence of master narratives, themselves the product of a modern ideology of science. Although presented as liberatory, Lyotard's invocations nevertheless reinstall the scientific as a primary mode of cultural explanation.

13. Bacon constructs inquiry into nature as coterminous with the agenda of royal power and observation as a subspecies of surveillance. *The Great Instauration* compares natural philosophers to "emissaries of princes" whose "diligence in exploring and unravelling plots and civil secrets" entitles them to be well paid (4:287).

14. The word appears in the unique edition of *The New Atlantis* to bear the date 1626 (STC 1168, copy 1, Folger Library), which may have been Bacon's proofcopy; all 1627 editions seen through the press by William Rawley also refer to the text as "unperfected."

15. Hodgen, *Early Anthropology,* 221–230.

16. See Samuel Purchas, *Hakluytus Posthumus, or Purchas His Pilgrimes* (London, 1625); *Purchas His Pilgrimage* (London, 1614).

17. The mystification of process typified by the circular affirmation of faith in *The New Atlantis* may also be found in the records that Bacon keeps of his material investigations. See, for example, his *Inquiry Concerning the Magnet,* 5:401–405.

18. My implicit definition of the theatrical may seem simplistic here, for it neglects the extent to which the audience constitutes the meaning of a performance. In the revelation scene, however, the drama merges conceptually with the masque: although the audience is located in theatrical space, it is there to witness an action brought about by a privileged viewer.

19. See Stephen Orgel, *The Illusion of Power: Political Theatre in the English Renaissance* (Berkeley: University of California Press, 1975).

20. Accommodating scientific inquiry to the demands of piety in the Renaissance involves renegotiating the boundaries of licit human knowledge within a Scriptural tradition of containment. For a discussion of that tradition, and the place of the figure of Icarus within it, see Carlo Ginzburg, "High and Low: The Theme of Forbidden Knowledge in the Sixteenth and Seventeenth Centuries," *Past and Present* 73 (1976): 29–41. *Paradise Lost* (8.66–411) marks the conflicts between earthbound piety and loftier flights of philosophical inquiry, since Raphael's advice to Adam to be "lowly wise" (8.173) follows his hypothetical discussion of Copernicanism in a poem circumscribed by the Icarus-like figure of Satan. See chapter 4.

21. See David Bergeron, *English Civic Pageantry, 1558–1642* (London: Edward Arnold, 1971); and Graham Parry, *The Golden Age Restor'd: The Culture of the Stuart Court* (Manchester: Manchester University Press, 1981).

22. The identification of James with Solomon was pervasive in Jacobean culture. Thus, in the oration delivered at James's funeral, Bishop Williams of Lincoln presented a lengthy set of parallels between James's virtues and those of the Old Testament king. See Robert Ashton, ed., *James I by His Contemporaries* (London: Hutchinson, 1969), 19–21. The frequency of quotations from, and references to, Solomon throughout Bacon's writings suggests a pervasive hyperconsciousness of monarchical regard typical of courtly literary productions under Elizabeth, as Louis Montrose, Leah Marcus, and Roy Strong have argued. For instance, even in the ostensibly private *Essays* (which, as "studies and contemplations" for which Bacon is "fittest," are contrasted with Bacon's official duties [6:523–524]), the name Solomon appears more often with each successive publication. In the 1598 edition there are no references to the Old Testament king; in the 1612 edition Solomon is mentioned at least a dozen times in nine essays; and by the final version of 1625 there are seventeen references in eleven essays. In fact, the name of Solomon occurs more often than that of any other historical, biblical, or classical figure.

23. See *Basilikon Doron*, in *The Political Works of James I*, ed. C. H. McIlwain (Cambridge, Mass.: Harvard University Press, 1918): "A king is as one set on a stage, whose smallest actions and gestures all the people gazingly behold" (43). The precariousness implicit in such display has been pointed out by Stephen Orgel in comparing the first private edition of the text with the second; rather than "stage," the earlier, emended, text reads "skaffold." See "Making Greatness Familiar," in *The Power of Forms in the English Renaissance,* ed. Stephen Greenblatt (Norman, Okla.: Pilgrim Books, 1982), 41–48. For related discussions of display and its ambiguous instantiation of monarchical power, see Christopher Pye, "The Sov-

ereign, the Theater, and the Kingdome of Darknesse," *Representations* 8 (1984): 85–106; and Francis Barker, *The Tremulous Private Body: Essays on Subjection* (London: Methuen, 1984).

24. Leonard Tennenhouse, *Power on Display: The Politics of Shakespeare's Genres* (New York: Methuen, 1986), esp. 147–186, has offered a related argument about Jacobean "disguised monarch" plays. In *Measure for Measure,* for example, the volitional absence of the ruler leads to an instructive disorder in the state, which can be resolved only by strategies of containment recasting Jacobean patriarchialism as legitimacy, the familial reproduction of right governance. In *The New Atlantis,* however, the fruit of paternity is the natural philosophy of the Salomonic Fathers, and its operations have more in common with the contemporaneously emerging logic of accumulation than with the maintenance of central political authority.

25. The father whose celebration is taking place receives a king's charter, sealed with an image of the king (3:149). As represented within *The New Atlantis,* then, the monarch is but the shadow of a shadow, at a second remove from the plane of representation delimited by the text.

26. Jonathan Goldberg, *James I and the Politics of Literature* (Baltimore: Johns Hopkins University Press, 1983), 149 -163. The proprietariness of the monarchical gaze—that the king owns all that he sees because he sees it—is dismantled by Bacon's text, even as the mechanism of evidentiary control, possessed by an elite not necessarily identifiable with a hereditary aristocracy, is reaffirmed; hence the possibility of scientific colonialism. For a connection between the primacy of visual evidence and conceptual othering in anthropology, see Johannes Fabian, *Time and the Other: How Anthropology Makes Its Object* (New York: Columbia University Press, 1983), esp. "The Other and the Eye," 105–141.

27. This argument is congruent with that advanced by Margaret Jacob, *The Radical Enlightenment* (London: Allen and Unwin, 1981), esp. 65–73. The Royal Society, as Jacob suggests, restored the "correct" reading of Bacon and of royal prerogative in science. See also Charles Webster, *The Great Instauration: Science, Medicine, and Reform, 1626–1660* (New York: Holmes and Meier, 1976); and Michael Hunter, *Science and Society in Restoration England* (Cambridge: Cambridge University Press, 1981).

28. For a detailed discussion not only of the influence of fifteenth- and sixteenth-century humanism on education, but of its transformation during the Renaissance into elite technical training, see Anthony Grafton and Lisa Jardine, *From Humanism to the Humanities* (Cambridge, Mass.: Harvard University Press, 1986).

29. As I have suggested, the formation of cultural hierarchies and disciplinary boundaries is intimately related to the demonizing—or feminizing, which is much

the same thing—of other segments of culture. Bacon, for instance, links the humanist love of words with a debased eroticism by comparing it to Pygmalion's frenzy (3:284). See also Joan Scott, "Gender: A Useful Category of Historical Analysis," in *Coming to Terms: Feminism, Theory, Politics,* ed. Elizabeth Weed (New York: Routledge, 1989), 81–100; and chapter 2 above.

30. Quoted in Edward D. Neill, *English Colonization of America During the Seventeenth Century* (London: Strahan, 1871), 28.

31. In "Romance and the Novel," Margaret Doody suggests that a new trope needs to be defined: the breaking, which she calls *rexis,* that provides the seismic crack to launch narrative into being. (Unpublished paper presented at 1980 MLA convention, New York.) Romances often call for shipwrecks, but all forms of fiction, Doody argues, may perhaps begin by thematizing ruptures. To move Doody's formalist argument into the realm of the historical, Bacon's becalmed ship signals its difference from a convention widely associated with the literary representation of colonialism in the English Renaissance from *The Tempest* to those documents concerning the Virginia Company's shipwreck off the Bermudas.

32. Quoted in Tzvetan Todorov, *The Conquest of America: The Question of the Other,* trans. Richard Howard (New York: Harper and Row, 1984), 17.

33. The parenthetical qualifiers are also the fruits of surveillance: the ubiquitousness of watching eyes leads to narrative paranoia and a concomitant need to authorize meaning according to a system of restriction opened up only at the conclusion. This point owes much to discussion with Harry Berger, Jr.

34. For Sidney's *Apology* as the aesthetics of an embattled humanist elite, see Terry Eagleton, *Criticism and Ideology: A Study in Marxist Literary History* (London: Verso, 1978), 18–19. Michael McKeon more amply treats the general point concerning the inadequacy of aristocratic literary forms for encompassing emergent ideologies and structures of feeling. See *The Origins of the English Novel, 1600–1740* (Baltimore: Johns Hopkins University Press, 1987).

35. See Brian Vickers's introduction to *Occult and Scientific Mentalities in the Renaissance* (Cambridge: Cambridge University Press, 1984), 1–55; in it he discusses Robin Horton's distinction between open and closed systems of knowledge (34–35). For a critique, see chapter 2 above.

36. Luce Irigaray, "Is the Subject of Science Sexed?," trans. Carol Mastrangelo Bové, in Tuana, ed., *Feminism and Science,* 58–68.

37. For a further exploration of the specific gender politics of Baconian science, see Merchant, *The Death of Nature.*

38. See Michel de Certeau, *The Writing of History,* xxv–xxvi; Louis Montrose, "The Work of Gender in the Discourse of Discovery," *Representations* 33 (1991): 1–41.

4 The Prosthetic Milton

1. Unless otherwise noted, all quotations from Milton's texts are taken from *John Milton: Complete Poems and Major Prose,* ed. Merritt Y. Hughes (New York: Macmillan, 1957).

2. For a model, I have in mind John Guillory's lucid discussion of the topic in *Samson Agonistes,* "Dalilah's House: *Samson Agonistes* and the Sexual Division of Labor," in *Rewriting the Renaissance: The Discourses of Sexual Difference in Early Modern Europe,* ed. Margaret Ferguson, Maureen Quilligan, and Nancy Vickers (Chicago: University of Chicago Press, 1986), 106–122. See also Guillory's "From the Superfluous to the Supernumerary: Reading Gender into *Paradise Lost,*" in *Soliciting Interpretation: Literary Theory and Seventeenth-Century Poetry,* ed. Elizabeth Harvey and Katharine Eisaman Maus (Chicago: University of Chicago Press, 1990), 68–88.

3. This argument crucially depends on an argument advanced by Nancy Armstrong and Leonard Tennenhouse in which they read the emergence of early modern (which is, for them, liberal humanist) concepts of man as a laboring body in *Paradise Lost.* See "The Work of Literature," in *The Imaginary Puritan: Literature, Intellectual Labor, and the Origins of Personal Life* (Berkeley: University of California Press, 1992), 89–113. As will become clear, however, I depart from their contention that gender as a category of analysis emerges as a consequence when liberal humanism emerges from Renaissance textual practice. While the oppositions between two genders is clearly reinflected by such cultural shifts—witness the normative heterosexuality of scientific modeling—to connect "gender" as such to these shifts is in effect to argue for the naturalness of sex roles in the Renaissance and before.

4. Gaston Bachelard, *The New Scientific Spirit,* trans. Arthur Goldhammer (Boston: Beacon Press, 1984), 13.

5. Marjorie Hope Nicolson, "Milton and the Telescope," in *Science and Imagination* (Ithaca, N.Y.: Cornell University Press, 1956), 81.

6. Harinder Singh Marjara, *Contemplation of Created Things: Science in* Paradise Lost (Toronto: University of Toronto Press, 1992), 85; see also Kester Svendsen, *Milton and Science* (Cambridge, Mass.: Harvard University Press, 1956).

7. Michel Foucault, *The History of Sexuality, Part I: An Introduction,* trans. Robert Hurley (New York: Random House, 1978); Thomas Laqueur, *Making Sex: Body and Gender from the Greeks to Freud* (Cambridge, Mass.: Harvard University Press, 1990).

8. Foucault, *History of Sexuality, Part I,* 156.

9. This fantasy of loss seems constitutive of Foucault's project. As I have sug-

gested elsewhere in this study, Foucault effectively leaves the sixteenth and seventeenth centuries as the uninterrogated origin or primal scene for the emergence of modern epistemes. The exception is *The Order of Things,* but even there, Foucault suggests that the analogic episteme supplanted by classical taxonomies may survive in what we now name as "literature"—a recourse to the aesthetic that precisely ties in with his discussion of the premodern body in terms of the pastoral. See my discussion in chapter 1.

10. Donna Haraway, "A Cyborg Manifesto: Science, Technology, and Socialist-Feminism in the Late Twentieth Century," printed most recently in Haraway's collection *Simians, Cyborgs, and Women: The Reinvention of Nature* (New York: Routledge, 1991), 149–181.

11. It is difficult to find language to discuss what is, after all, a preformation. I am in effect imagining a theoretical space, or an allegory of a nascent cultural division of labor, that stands somewhere between the liberal art of the humanists and the accomplished institutions of the Enlightenment and after. For the division between "fact" and "fiction," see chapter 1 above, and Bernard Weinberg, *A History of Literary Criticism in the Italian Renaissance* (Chicago: University of Chicago Press, 1961); Barbara J. Shapiro, *Probability and Certainty in Seventeenth-Century England: A Study of the Relationships Between Natural Science, Religion, History, Law, and Literature* (Princeton, N.J.: Princeton University Press, 1983); Gayle Ormiston and Raphael Sassower, *Narrative Experiments: The Discursive Authority of Science and Technology* (Minneapolis: University of Minnesota Press, 1989), 3–35. For related accounts of the origins of disciplines and institutions, see Michael McKeon, *The Origins of the English Novel, 1600–1740* (Baltimore: Johns Hopkins University Press, 1987); John Barrell, *The Birth of Pandora and the Division of Knowledge* (Philadelphia: University of Pennsylvania Press, 1992); Michael Hunter, *Science and Society in Restoration England* (Cambridge: Cambridge University Press, 1981); and Timothy J. Reiss, *The Meaning of Literature* (Ithaca, N.Y.: Cornell University Press, 1992).

12. John Guillory has made a similar argument concerning the references to the telescope and the irruption of "history" into *Paradise Lost.* See *Poetic Authority: Spenser, Milton, and Literary History* (New York: Columbia University Press, 1983), 156–167. The readiness with which Guillory moves from the instrument to the astronomer presumed to be its inventor, its point of origin, signals the point where our arguments diverge. Whether Galileo is "the return of the [historical] repressed" *or* "a cryptic self-portrait" (161)—Guillory moves from one position to the other—the movement backward from apparatus to embodiment insists on the retrievability of human agency. History does not write the body out of existence.

13. William Riley Parker, *Milton: A Biography* (Oxford: Clarendon Press, 1968), 1:178–179.

14. Guillory, *Poetic Authority,* 156.

15. Guillory, "From the Superfluous to the Supernumerary," 85.

16. See, for example, the editions of Merritt Hughes, Alastair Fowler, Edward LeComte, and Douglas Bush. I should note that I owe much of the argument that follows to discussions with Zofia Burr.

17. Galileo was not the actual inventor of the telescope, although he made his own telescope based on a verbal description and may well have improved upon what was described to him. In 1608 (before Galileo began his celestial observations) two Dutchmen, Jan Lippershey and James Metius, filed for the "exclusive manufacture and sale of the instruments." See Nicolson, *Science and Imagination,* 9–15.

18. See Timothy Reiss, *The Discourse of Modernism* (Ithaca, N.Y.: Cornell University Press, 1982), 24–34.

19. The word "telescope" had come into English by then, although generally in the Latin and Italian forms used on the Continent by Galileo, Porta, and Kepler. Thus, the OED cites Bainbridge in 1619 on "the *Telescopium* or Trunke-spectacle," and Boyle's familiarity with "Galileo's opticke Glasses . . . [or] Telescopioes" in 1648. The name was not wholly stabilized in the seventeenth century; depending on the language of citation, there are references to the "perspicillium," the "cannochiale," and, as Milton shows, the "optic glass."

20. Nicolson, *Science and Imagination,* 89–90.

21. I have in mind Foucault's *Discipline and Punish: The Birth of the Prison,* trans. Alan Sheridan (New York: Random House, 1978), his most lucid exposition of the productive effects of technology on the embodied subject, who, as Althusser has argued, must be shaped so as to "work by [himself]." See also Louis Althusser, "Ideology and Ideological State Apparatuses," in *Lenin and Philosophy and Other Essays,* trans. Ben Brewster (New York: Monthly Review Press, 1971), 127–186.

22. Raymond Williams, *The Country and the City* (New York: Oxford University Press, 1973).

23. See Timothy Hampton, *Writing from History: The Rhetoric of Exemplarity in Renaissance Literature* (Ithaca, N.Y.: Cornell University Press, 1990); Victoria Kahn, "Habermas, Machiavelli, and the Humanist Critique of Ideology," *PMLA* 105 (1990): 464–476.

24. Stephanie Jed, *Chaste Thinking: The Rape of Lucretia and the Birth of Humanism* (Bloomington: Indiana University Press, 1989).

25. This summary analysis owes much to Catherine Belsey's provocative and economical insights about *Paradise Lost* as a production in history. See Belsey, *John Milton: Language, Gender, Power* (Oxford: Basil Blackwell, 1988), esp. 35–36.

26. See Foucault, *The Order of Things,* trans. of *Les Mots et les Choses* (New York: Random House, 1970), 386–387.

27. Of course, the claim to be free from ideology is an ideological claim in itself; Galileo's invention of a textual space "beyond" the constraints of papal dogma must be read specifically.

28. Belsey, *John Milton,* 74.

29. The war in heaven, which was taken in the eighteenth century to be sublimely Homeric, now can seem embarrassing in its lack of epic decorum—although it can be rescued from the charge through a bit of special pleading: the heroism of arms is *meant* to be ludicrous in a poem that praises "the better fortitude / Of patience and heroic martyrdom" (9.31–32). It seems to me just as useful to read this parodic element in Bakhtinian terms—not just as a carnivalesque rebellion of underlings, but as a marker of the incipient *historical* untenability of the epic, given the emergence of the novel and the subjectivities it models. See M. M. Bakhtin, "Epic and Novel," in *The Dialogic Imagination: Four Essays,* ed. Michael Holquist, trans. Caryl Emerson and Holquist (Austin: University of Texas Press, 1981), 3–40.

30. See Helen Gardner, "Milton's 'Satan' and the Theme of Damnation in Elizabethan Tragedy," in *Milton: Modern Essays in Criticism,* ed. Arthur E. Barker (New York: Oxford University Press, 1965), 205–217; B. Rajan, *Paradise Lost and the Seventeenth-Century Reader* (Ann Arbor: University of Michigan Press, 1967), 93–107; Kenneth Gross, "Satan and the Romantic Satan: A Notebook," in *Remembering Milton: Essays on the Texts and Traditions,* ed. Mary Nyquist and Margaret W. Ferguson (New York: Methuen, 1987), 318–341.

31. While the word "effeminate" is not used to describe Adam's passion for Eve, his words leave no doubt that such love is a reversal of natural hierarchy; see 8.540–594. For my general sense of the relationship between effeminacy and married love, I am indebted to Gary Spear's related argument about Charles I, Henrietta Maria, and *Samson Agonistes.* See his "Nation and Effemination" (unpublished paper read at the Folger Shakespeare Library, Oct. 1992). For a related argument, see Regina Schwartz, "Rethinking Voyeurism and Patriarchy: The Case of *Paradise Lost,*" *Representations* 34 (1991): 99. For the basis of "effeminacy" in theories of the body, see Laqueur, *Making Sex,* 122–128.

32. Laqueur, *Making Sex,* 149ff.

33. Laqueur, *Making Sex,* 153. It is worth noting that Laqueur, for all his useful skepticism about the empirical standpoint of scientific knowledge, does not avoid replicating its logic of visual evidence—not only in the fact of his illustrations, but in the very decision to write a history of "sex" as embodied.

34. Schwartz, "Rethinking Voyeurism and Patriarchy."

35. Fredric Jameson, *Signatures of the Visible* (New York: Routledge, 1992), 1; emphasis Jameson's. It is curious that, given the breadth of his claim about the

libidinality of all visual culture, Jameson spends the book analyzing only the film culture of the later twentieth century—thereby betraying the historicity of his claim of essence.

36. Linda Williams, *Hardcore: Power, Pleasure, and the Frenzy of the "Visible"* (Berkeley: University of California Press, 1989).

37. Of course, I have in mind Foucault's analysis of the power-knowledge-pleasure nexus, a concept that also informs Linda Williams's discussion of *scientia sexualis* and modern pornography. Foucault, *The History of Sexuality, Part I,* esp. 57–73.

38. Armstrong and Tennenhouse, *The Imaginary Puritan,* 103–113.

39. Guillory makes a similar point regarding the unoriginality of the image. See "Reading Gender," 76–77.

40. Thomas Harriot, *A Briefe and True Report of the New Found Land of Virginia* (1590), intro. Paul Hulton (New York: Dover, 1972).

41. Carlo Ginzburg, "High and Low: The Theme of Forbidden Knowledge in the Sixteenth and Seventeenth Centuries," *Past and Present* 73 (1976): 28–41.

42. Samuel Purchas, *Purchas His Pilgrimage* (London, 1614), 912. For a case study of just such conversion attempts and the hybrid cultural forms they produce, see Jonathan Spence, *The Memory Palace of Matteo Ricci* (New York: Penguin, 1984).

43. See *Areopagitica,* 723–725, 737–738, in Hughes's edition.

44. Hans Vaihinger, *The Philosophy of 'As-If,'* trans. C. K. Ogden (New York: Harcourt Brace, 1924).

45. Michel de Certeau, *The Writing of History,* trans. Tom Conley (New York: Columbia University Press, 1988), 3–6.

5 Galileo, "Literature," and the Generation of Scientific Universals

1. See Ian Maclean, *The Renaissance Notion of Woman: A Study in the Fortunes of Scholasticism and Medical Science in European Intellectual Life* (Cambridge: Cambridge University Press, 1980); Thomas Laqueur, *Making Sex: Body and Gender from the Greeks to Freud* (Cambridge, Mass.: Harvard University Press, 1990).

2. However, I have in most cases chosen to cite Galileo in his earliest English translation, that of Thomas Salusbury, published in 1661, as well as in Italian. This translation has manifold advantages: it reproduces the seventeenth-century cultural field; it provides a specifically English view of the papal proceedings against Galileo; finally, it avoids the technical diction of physics and astronomy used by most later translators, which give the misleading impression that Galileo is writing within an established discipline with a conventional vocabulary. Salus-

bury was in exile from England during the Protectorate; Jean Dietz Moss speculates that that fact may account for his "more than usual deference" to the Catholic position. For Galileo's dissemination in seventeenth-century England, see Moss, "Galileo Interpreted for Englishmen," in *Novelties in the Heavens: Rhetoric and Science in the Copernican Controversy* (Chicago: University of Chicago Press, 1993), 301–329. This book, many of whose observations on Galileo dovetail with my own, was published too late for me to make substantive use of it. Briefly, however, Moss pursues a more conventionally historical line in considering "rhetoric" to be the operative term of analysis in furthering an already existent "science."

3. Maurice A. Finocchiaro, *Galileo and the Art of Reasoning* (Boston Studies in the Philosophy of Science, vol. 61) (Dordrecht: Reidel, 1980); Jean Dietz Moss, "Galileo's *Letter to Christina:* Some Rhetorical Considerations," *Renaissance Quarterly* 36 (1983): 547–576; "The Rhetoric of Proof in Galileo's Writings on the Copernican System," in *Reinterpreting Galileo,* ed. William A. Wallace (Studies in Philosophy and the History of Philosophy, vol. 15) (Washington, D.C.: Catholic University Press of America, 1986), 179–204; see also Moss's *Novelties in the Heavens.*

4. Karl von Gebler, *Galileo Galilei and the Roman Curia,* trans. Mrs. George Sturge (London: C. Kegan Paul, 1879), 23–24. Gebler gives no date for the correspondence with Kepler. The letter of 30 July 1610, sent by Galileo to Belisario Vinta, secretary of state to Cosimo de Medici (Letter 370 of the National Edition) does speak in straightforward Italian about the observation that Saturn is a triple star and asks Vinta not to disseminate the observation. See *Le Opere di Galileo,* ed. Antonio Favaro (Florence: G. Barbera, 1930–), X. 409–410.

5. Letter 427, to Giuliano de Medici, 13 Nov. 1610; Gebler, 24. The Italian reads: "Ma passando ad altro, già che il S. Keplero ha in questa sua ultima Narrazione stampate le lettere che io mandai a V. S. Ill[ustrissi]ma trasposte, venendomi anco significato come S. M. ne desidera il senso. . . . Le lettere dunque, combinate nel loro vero senso, dicono così: Altissimum . . ." (*Opere,* X. 474). To be precise, the code as reprinted has an N, where another M is needed.

6. Gebler, *Galileo Galilei and the Roman Curia,* 23–25.

7. For a study of the relationship between the nascent humanist curricula and the development of a politically influential elite, see Anthony Grafton and Lisa Jardine, *From Humanism to the Humanities* (Cambridge, Mass.: Harvard University Press, 1986). Donald R. Kelley has argued that, while Renaissance humanism as strictly defined should be confined to the fifteenth and early sixteenth centuries, the term has wider value as a historical generalization—especially in connection with the subsequent divisions in cultural labor that I am concerned to map out. See Kelley, *Renaissance Humanism* (Boston: Twayne, 1991), 111–135.

8. See, for instance, Marjorie Hope Nicolson, *The Breaking of the Circle: Studies in the Effect of the 'New Science' on Seventeenth-Century Poetry* (New York: Columbia University Press, 1960); C. P. Snow, *The Two Cultures and A Second Look: An Expanded Version of the Two Cultures and the Scientific Revolution* (Cambridge: Cambridge University Press, 1964); Jacob Bronowski, *Science and Human Values* (New York: Harper and Row, 1965). Even Jean-François Lyotard's more recent study of postmodernity and grand narratives works within a binary model of culture. See *The Postmodern Condition: A Report on Knowledge,* trans. Geoff Bennington and Brian Massumi (Minneapolis: University of Minnesota Press, 1984).

9. The English translations of Galileo's *Dialogo* are taken from Thomas Salusbury's seventeenth-century English translation, *Galileo Galileus His System of the World,* contained in the First Part of *Mathematical Collections and Translations: The First Tome; in Two Parts* (London: William Leybourn, 1661), 92. Further page references are provided in the text. "Centone" is given this gloss in the margins of Salusbury's translation: "books composed of many fragments of verses collected out of the Poets"; they appear to resemble commonplace books.

10. The Italian can be found in VII. 134–135.

11. For the early history of this topos, see E. R. Curtius, *European Literature and the Latin Middle Ages,* trans. Willard Trask (Princeton, N.J.: Princeton University Press, 1953), 319–326.

12. Chandra Mukerji, *From Graven Images: Patterns of Modern Materialism* (New York: Columbia University Press, 1983), 131. Elizabeth L. Eisenstein offers a useful historical distinction between the sense in which the "book of nature" is understood by the medieval writers whom Curtius surveys and the sense that the phrase carried for Galileo and others. See *The Printing Press as an Agent of Change: Communications and Cultural Transformations in Early Modern Europe* (Cambridge: Cambridge University Press, 1979), 453ff.

13. See Kelley, *Renaissance Humanism,* 2–4; 130–132.

14. Gebler, *Galileo Galilei and the Roman Curia,* 100.

15. Special injunction issued by Cardinal Robert Bellarmine, 26 Feb. 1616. The translation by Maurice A. Finocchiaro appears in his invaluable *The Galileo Affair: A Documentary History* (Berkeley: University of California Press, 1989), 147.

16. This would appear to explain why Galileo continued his investigations with impunity until the *Dialogue*'s suppression immediately after it was printed.

17. Michel Serres, "Platonic Dialogue," in *Hermes: Literature, Science, Philosophy,* ed. Josue Harari and David F. Bell (Baltimore: Johns Hopkins University Press, 1982), 65–70.

18. *Dialogue Concerning the Two Chief World Systems—Ptolemaic and Copernican,*

2d ed., trans. and ed. Stillman Drake (Berkeley: University of California Press, 1967), 162.

19. Leo Strauss, *Persecution and the Art of Writing* (1952; Chicago: University of Chicago Press, 1988). See also Annabel Patterson, *Censorship and Interpretation: The Conditions of Writing and Reading in Early Modern England* (Madison: University of Wisconsin Press, 1984).

20. The English translations appear in *Discoveries and Opinions of Galileo*, trans. and ed. Stillman Drake (New York: Doubleday/Anchor, 1957), 199. Compare this passage to one taken from the *Dialogue*'s Second Day, when Salviati speaks of "Chymists" who find nothing but the secret of making gold in texts: "but that they might transmit the secret to posterity without discovering it to the vulgar, they contrived . . . to conceal the same under several maskes; and it would make one merry to hear their comments upon the ancient *Poets*, finding out the important misteries, which lie hid under their Fables; and the signification of the Loves of the *Moon*, and her descending to the Earth for *Endimion;* her displeasure against *Actaeon*, and what was meant by *Jupiters* turning himself into a showre of *Gold* . . ." (Salusbury, *Mathematical Collections*, 92).

21. Michel de Certeau, *The Writing of History*, trans. Tom Conley (New York: Columbia University Press, 1988), 3.

22. "*Fantasia*," whose connotations in Italian are overwhelmingly literary, is Galileo's word for what is now generally termed a hypothesis.

23. The most polemical account may be Giorgio di Santillana's *The Crime of Galileo* (Chicago: University of Chicago Press, 1955).

24. William D. Montalbano, "Vatican Finds Galileo 'Not Guilty': Pope Admits Error in Rejecting Theory," *Washington Post*, 1 Nov. 1992, A40.

25. John Paul II's citation from *Gaudium et Spes* on the necessary separation between science and the church is quoted in Ernan McMullin's foreword to Robert S. Westfall, *Essays on the Trial of Galileo, Studi Galileiani* no. 5 (Vatican City: Vatican Observatory Publications, 1989), v–vi.

26. Pietro Redondi, *Galileo Heretic*, trans. Raymond Rosenthal (Princeton, N.J.: Princeton University Press, 1987). For one criticism of Redondi's argument, see Richard S. Westfall, "*Galileo Heretic:* Problems, as They Appear to Me, with Redondi's Book," in *Essays on the Trial of Galileo*, 84–99.

27. Peter Armour, "Galileo and the Crisis in Italian Literature," in *Collected Essays on Italian Language and Literature Presented to Kathleen Speight*, ed. Giovanni Aquilecchia et al. (New York: Barnes and Noble, 1971), 144.

28. Stillman Drake, foreword to Jerome J. Langford, *Galileo, Science, and the Church*, 3d ed. (Ann Arbor: University of Michigan Press, 1992), ix.

29. In the course of an essay evaluating the status of occultism in histories of science, Paolo Rossi has written scathingly (and rather narrowly) against the

"mythical portraits" produced within much recent history of science, where "Bacon or Galileo . . . have become 'symbols' for certain points of view, 'masks' for inductivism or hypothetico-deductivism" (247). See "Hermeticism, Rationality, and the Scientific Revolution," in *Reason, Experiment, and Mysticism in the Scientific Revolution,* ed. M. L. Righini Bonelli and William R. Shea (New York: Science History, 1975), 247–273. My different predication should make clear why I find Rossi's position (and his faith in a pure historiography of science that has a lot in common with its object) other than wholly convincing.

30. Michel Foucault, "What Is an Author?" in *Language, Counter-Memory, Practice,* ed. Donald Bouchard; trans. Bouchard and Sherry Simon (Ithaca, N.Y.: Cornell University Press, 1977), 135–136.

31. See Hans Blumenberg, *The Legitimacy of the Modern Age,* trans. Robert M. Wallace (Cambridge, Mass.: MIT Press, 1983), which accepts that modern science emerges with the controversies around the Copernican hypothesis. A more sustained account can be found in Blumenberg's *The Genesis of the Copernican World,* trans. Robert M. Wallace (Cambridge, Mass.: MIT Press, 1987). For a collection that problematizes the "revolution" in knowledge of the seventeenth century, see *Reappraisals of the Scientific Revolution,* ed. David C. Lindberg and Robert S. Westman (Cambridge: Cambridge University Press, 1990).

32. Einstein would be a more recent example. Of course, I am overstating the case somewhat. To be sure, Foucault also mentions Newton, for example, as well as Cuvier—although, significantly, not Descartes. However, Foucault does return three times to Galileo as a negative example.

33. References to the Copernican "revolution" are, of course, pervasive in historiography. See, for but one instance, Thomas S. Kuhn's *The Copernican Revolution: Planetary Astronomy in the Development of Western Thought* (Cambridge, Mass.: Harvard University Press, 1957). For a defense of the model against Kuhn's later work, see I. Bernard Cohen, *Revolution in Science* (Cambridge, Mass.: Harvard University Press, 1985). In speaking of Kuhn's "later work," and of the concept of "normal science" in general, I refer to the widely influential *The Structure of Scientific Revolutions,* 2d ed. (Chicago: University of Chicago Press, 1970). Even in this account of a nonlinear, non-"progressive" science, however, the models of "paradigm" and "consensus" that define normality within science are typical of an institution produced within and by a mature, dominant episteme. For further critiques of the Kuhnian "paradigm shift," see Imre Lakatos and Alan Musgrave, eds., *Criticism and the Growth of Knowledge* (Cambridge: Cambridge University Press, 1970).

34. The Italian passages are taken from the National Edition of *Le Opere di Galileo,* ed. Antonio Favaro (Florence: Barbera, 1933), esp. vol. 7, *I Due Massimi Sistemi Del Mondo.*

35. See Drake, *Dialogue,* 467, notes to p. 7.

36. Drake disputes that Simplicio can be directly identified with Pope Urban VIII, deeming such a potential representation "a preposterous piece of insolence serving no purpose except malice" (468). However, the defeated Simplicio ends the *Dialogue* by citing Urban's refusal to limit divine power through logical and mathematical exigency; if such ventriloquism is not "malice," it may well be provocation. The letter to the reader thus identifies him: "un filosofo peripatetico, al quale pareva che niuna cosa sotasse maggiormente per l'intelligenza del vero, che la fama acquistata nell'interpretazioni Aristoteliche" (VII. 31).

37. Jerome J. Langford does not believe, as other historians have done, that Galileo knew the tidal argument was inadequate. See *Galileo, Science, and the Church*, 126–127, 197–198.

38. See Langford, *Galileo, Science, and the Church*, 113–136.

39. Santillana, *Crime of Galileo*, passim.

40. Serres, "Platonic Dialogue," 65–70.

41. For example, when Salviati argues for the counterintuitive motion of a ball dropped from the hand of a moving rider on horseback, Simplicio expostulates: "But for Gods sake, if it move transversely, how is it that I behold it to move directly and perpendicularly? This is no better than the denial of manifest sense; and if we may not believe sense, at what other door shall we enter into the disquisitions of Philosophy?" (Salusbury, *Mathematical Collections,* 151).

42. Victoria Kahn, "Habermas, Machiavelli, and the Humanist Critique of Ideology" *PMLA* 105, 3 (1990): 464–476.

43. See Paul Feyerabend, *Against Method* (London: Verso, 1975). An article by Boyce Rensburger, "Optics: Galileo's 'Perfect' Lens of 1609," in the *Washington Post,* 13 July 1992, A2, however, suggests that those of Galileo's lenses preserved by the Medici family and recently tested were surprisingly accurate, even by today's standards.

44. Thomas Kuhn, "A Function for Thought-Experiments," in *The Essential Tension: Selected Studies in Scientific Tradition and Change* (Chicago: University of Chicago Press, 1977), 240–265.

45. Sagredo is interrogating Simplicio on the opposition he demands between perfect and imperfect forms: is it, for instance, more difficult to make a material sphere absolutely perfect? What happens with other Platonically inflected representations, as with the statue of a horse? Is its form more or less irregular than those obtained by breaking a rock with a hammer?

Simp: "So it should be."

Sag.: "But tell me; that figure what ever it is which the stone hath, it hath that same in perfection, or no?"

Simp.: "What it hath, it hath so perfectly, that nothing can be more exact."

Sag: "Then, if of figures that are irregular, and consequently hard to be procured, there are yet infinite which are most perfectly obteined, with what reason can it be said, that the most simple, and consequently the most easie of all [i.e., the sphere], is impossible to be procured?" (Salusbury, *Mathematical Collections,* 187).

46. Hans Vaihinger, *The Philosophy of 'As-If',* trans. C. K. Ogden (New York: Harcourt, Brace, 1924); see also Lon L. Fuller, *Legal Fictions* (Stanford, Calif.: Stanford University Press, 1967).

47. The allegory is not entirely far-fetched, given the equivalence argued for in the Italian Renaissance between poetry and painting and the complex interplay of the arts commemorated, for example, in Giambattista Marino's collection of poems, *La Galeria.* For an extended investigation of the former, see Rensselaer W. Lee, *Ut Pictura Poesis: The Humanistic Theory of Painting* (New York: Norton, 1967). For that matter, Galileo was himself interested in the *paragone,* the comparison between the mimetic capacity of the arts, as a letter to Lodovico Cigoli in 1612 makes clear. See Erwin Panofsky, *Galileo as a Critic of the Arts* (The Hague: M. Nijhoff, 1954).

48. The line "mille e mille versi" is not translated in Salusbury, whose translation appears in the body of the chapter. Nor is the ambiguity of the line adequately rendered by Drake's "in thousands of directions" (172).

49. Jonathan Goldberg, *Writing Matter: From the Hands of the English Renaissance* (Stanford, Calif.: Stanford University Press, 1990).

50. Louis Althusser, "Ideology and Ideological State Apparatuses," in *Lenin and Philosophy and Other Essays,* trans. Ben Brewster (New York: Monthly Review Press, 1971), esp. 166–182; Michel Foucault, *Discipline and Punish: The Birth of the Prison,* trans. Alan Sheridan (New York: Random House, 1977).

Conclusion

1. Gayatri Spivak, "Can the Subaltern Speak?" in *Marxism and the Interpretation of Cultures,* ed. Lawrence Grossberg and Cary Nelson (Urbana: University of Illinois Press, 1988), 271–313.

2. Michel Foucault and Gilles Deleuze, "Intellectuals and Power," in *Language, Counter-Memory, Practice: Selected Essays and Interviews,* ed. Donald F. Bouchard, trans. Bouchard and Sherry Simon (Ithaca, N.Y.: Cornell University Press, 1977), 205–217.

3. See Cornel West, "The Postmodern Crisis of the Black Intellectuals," in

Cultural Studies, ed. Lawrence Grossberg, Cary Nelson, and Paula Treichler (New York: Routledge, 1992), 689–705.

4. See Michel de Certeau, *Heterologies: Discourse on the Other,* trans. Brian Massumi (Minneapolis: University of Minneapolis Press, 1986), 185–192.

5. Michel Foucault, *Madness and Civilization: A History of Insanity in the Age of Reason,* trans. Richard Howard (New York: Random House, 1965), xi.

Works Cited

Alexander, Elizabeth. *The Venus Hottentot.* Charlottesville: University Press of Virginia, 1990.

Althusser, Louis. "Ideology and Ideological State Apparatuses." In *Lenin and Philosophy and Other Essays.* Trans. Ben Brewster. New York: Monthly Review Press, 1971, 127–186.

Armour, Peter. "Galileo and the Crisis in Italian Literature." In *Collected Essays on Italian Language and Literature Presented to Kathleen Speight.* Ed. Giovanni Aquilecchia et al. New York: Barnes and Noble, 1971, 144–169.

Armstrong, Nancy, and Leonard Tennenhouse. *The Imaginary Puritan: Literature, Intellectual Labor, and the Origins of Personal Life.* Berkeley: University of California Press, 1992.

Ashton, Robert, ed. *James I by His Contemporaries.* London: Hutchinson, 1969.

Bachelard, Gaston. *The New Scientific Spirit.* Trans. Arthur Goldhammer. Boston: Beacon Press, 1984.

Bacon, Francis. *Francis Bacon: A Selection of His Works.* Ed. Sidney Warhaft. New York: Odyssey Press, 1965.

———. *The Works of Francis Bacon.* Ed. James Spedding, Robert Leslie Ellis, and Douglas Denon Heath. New York: Garrett Press, 1968.

Bakhtin, M. M. "Epic and Novel." In *The Dialogic Imagination: Four Essays.* Ed. Michael Holquist; trans. Caryl Emerson and Holquist. Austin: University of Texas Press, 1981, 3–40.

Barker, Francis. *The Tremulous Private Body: Essays on Subjection.* London: Methuen, 1984.

Barrell, John. *The Birth of Pandora and the Division of Knowledge.* Philadelphia: University of Pennsylvania Press, 1992.

Baudrillard, Jean. "The Ecstasy of Communication." In *The Anti-Aesthetic: Essays on Postmodern Culture.* Ed. Hal Foster. Seattle: Bay Press, 1983, 126–134.

Belsey, Catherine. *John Milton: Language, Gender, Power.* Oxford: Basil Blackwell, 1988.

———. *The Subject of Tragedy: Identify and Difference in Renaissance Drama.* New York: Methuen, 1985.

Works Cited

Bender, John, and David E. Wellbery, eds. *The Ends of Rhetoric: History, Theory, Practice*. Stanford, Calif.: Stanford University Press, 1990.

Bergeron, David. *English Civic Pageantry, 1558–1642*. London: Edward Arnold, 1971.

Biagioli, Mario. *Galileo, Courtier: The Practice of Science in an Age of Absolutism*. Chicago: University of Chicago Press, 1993.

Blake, Ralph, Curt J. Ducasse, and Edward Madden. *Theories of Scientific Method: The Renaissance Through the Nineteenth Century*. Ed. Edward Madden. Seattle: University of Washington Press, 1960.

Blumenberg, Hans. *The Genesis of the Copernican World*. Trans. Robert M. Wallace. Cambridge, Mass.: MIT Press, 1987.

———. *The Legitimacy of the Modern Age*. Trans. Robert M. Wallace. Cambridge, Mass.: MIT Press, 1983.

Boesky, Amy. " 'Outlandish-Fruits': Commissioning Nature for the Museum of Man." *English Literary History* 58 (1991): 305–330.

Braudel, Fernand. *Civilization and Capitalism: 15th–18th Century*. Trans. Sian Reynolds. New York: Harper and Row, 1981.

Brewer, John, and Roy Porter, eds. *Consumption and the World of Goods*. New York: Routledge, 1993.

Bronowski, Jacob. *Science and Human Values*. New York: Harper and Row, 1965.

Brown, Paul. " 'This thing of darkness I acknowledge mine': *The Tempest* and the Discourse of Colonialism." In *Political Shakespeare: New Essays in Cultural Materialism*. Ed. Jonathan Dollimore and Alan Sinfield. Manchester: Manchester University Press, 1985, 48–71.

Bryson, Norman. "The Ideal and the Abject: Cindy Sherman's Historical Portraits." *Parkett* 29 (1991): 91–93.

Burke, Peter. *The Renaissance Sense of the Past*. New York: St. Martin's Press, 1969.

Busia, Abena. "Silencing Sycorax: On African Colonial Discourse and the Unvoiced Female." *Cultural Critique* 14 (1990): 81–104.

Cohen, I. Bernard. *Revolution in Science*. Cambridge, Mass.: Harvard University Press, 1985.

Copernicus, Nicolas. *Des Révolutions des Orbes Célestes*. Latin text trans. with intro. Alexandre Koyré. Paris: Blanchard, 1970.

———. *On the Revolution of the Heavenly Spheres*. Trans. with intro. A. M. Duncan. New York: Barnes and Noble, 1976.

Couliano, Ioan P. *Eros and Magic in the Renaissance*. Trans. Margaret Cook. Chicago: University of Chicago Press, 1987.

Crimp, Douglas. *On the Museum's Ruins*. Cambridge, Mass.: MIT Press, 1993.

Curtius, E. R. *European Literature and the Latin Middle Ages*. Trans. Willard Trask. Princeton, N.J.: Princeton University Press, 1953.

Works Cited

de Certeau, Michel. *Heterologies: Discourse on the Other.* Trans. Brian Massumi. Minneapolis: University of Minnesota Press, 1986.

———. *The Writing of History.* Intro. and trans. Tom Conley. New York: Columbia University Press, 1988.

di Santillana, Giorgio. *The Crime of Galileo.* Chicago: University of Chicago Press, 1955.

Dollimore, Jonathan. *Radical Tragedy: Religion, Ideology, and Power in the Drama of Shakespeare and His Contemporaries.* Chicago: University of Chicago Press, 1984.

Donaldson, Laura E. "The Miranda Complex: Colonialism and the Question of Feminist Reading." In *Decolonizing Feminisms: Race, Gender, and Empire-Building.* Chapel Hill: University of North Carolina Press, 1992, 13–31.

Donne, John. *Ignatius His Conclave.* Latin and English, ed. and annotated T. S. Healy, S.J. Oxford: Clarendon Press, 1969.

DuBois, Page. "Subjected Bodies, Science, and the State: Francis Bacon, Torturer." In *Body Politics: Disease, Desire, and the Family.* Ed. Michael Ryan and Avery Gordon. Boulder, Colo.: Westview Press, 1994, 175–191.

Eagleton, Terry. *Criticism and Ideology: A Study in Marxist Literary History.* London: Verso, 1978.

Eisenstein, Elizabeth. *The Printing Press as an Agent of Change: Communications and Cultural Transformations in Early Modern Europe.* Cambridge: Cambridge University Press, 1979.

Erasmus, Desiderius. *Dialogus Ciceronianus.* Ed. Pierre Mesnard. In *Opera Omnia,* v.1.2. Amsterdam: North Holland Publishing Co., 1971.

Fabian, Johannes. *Time and the Other: How Anthropology Makes Its Object.* New York: Columbia University Press, 1983.

Feyerabend, Paul. *Against Method.* London: Verso/NLB, 1972.

Finocchiaro, Maurice A. *Galileo and the Art of Reasoning.* (Boston Studies in the Philosophy of Science, vol. 61.) Dordrecht: Reidel, 1980.

———, ed. *The Galileo Affair: A Documentary History.* Berkeley: University of California Press, 1989.

Fish, Stanley. *Self-Consuming Artifacts.* Berkeley: University of California Press, 1972.

Foucault, Michel. *The Archaeology of Knowledge.* Trans. A. M. Sheridan Smith. New York: Harper and Row, 1972.

———. *The Birth of the Clinic: An Archaeology of Medical Perception.* Trans. A. M. Sheridan Smith. New York: Pantheon, 1973.

———. *Discipline and Punish: The Birth of the Prison.* Trans. Alan Sheridan. New York: Random House, 1978.

———. *The History of Sexuality, Part I: An Introduction.* Trans. Robert Hurley. New York: Random House, 1978.

———. Introduction to Georges Canguilhem, *The Normal and the Pathological.* Trans. Carolyn R. Fawcett, with Robert S. Cohen. New York: Zone Books, 1991.

———. *Madness and Civilization: A History of Insanity in the Age of Reason.* Trans. Richard Howard. New York: Random House, 1965.

———. *The Order of Things: An Archaeology of the Human Sciences.* Trans. of *Les Mots et Les Choses.* New York: Random House, 1970.

———. "Two Lectures." In *Power/Knowledge: Selected Interviews and Other Writings, 1972–1977.* Ed. Colin Gordon. New York: Pantheon, 1980, 78–108.

———. "What Is an Author?" In *Language, Counter-Memory, Practice: Selected Essays and Interviews.* Ed. Donald F. Bouchard. Trans. Donald F. Bouchard and Sherry Simon. Ithaca, N.Y.: Cornell University Press, 1977, 113–138.

———, and Gilles Deleuze. "Intellectuals and Power." In *Language, Counter-Memory, Practice: Selected Essays and Interviews.* Ed. Donald F. Bouchard. Trans. Donald F. Bouchard and Sherry Simon. Ithaca, N.Y.: Cornell University Press, 1977, 205–217.

Fuller, Lon L. *Legal Fictions.* Stanford, Calif.: Stanford University Press, 1967.

Galilei, Galileo. *Dialogue Concerning the Two Chief World Systems—Ptolemaic and Copernican.* 2d ed. Trans. and ed. Stillman Drake. Berkeley: University of California Press, 1967.

———. *Discoveries and Opinions of Galileo.* Trans. and ed. Stillman Drake. New York: Doubleday/Anchor, 1957.

———. *Le Opere di Galileo.* Ed. Antonio Favaro. Florence: G. Barbera, 1930.

Gardner, Helen. "Milton's 'Satan' and the Theme of Damnation in Elizabethan Tragedy." In *Milton: Modern Essays in Criticism.* Ed. Arthur E. Barker. New York: Oxford University Press, 1965, 205–217.

Garin, Eugenio. *Italian Humanism: Philosophy and Civic Life in the Renaissance.* Trans. Peter Munz. New York: Harper and Row, 1965.

———. *Science and Civic Life in the Italian Renaissance.* Trans. Peter Munz. New York: Doubleday, 1969.

Gebler, Karl von. *Galileo Galilei and the Roman Curia.* Trans. Mrs. George Sturge. London: C. Kegan Paul, 1879.

Gilman, Sander. "Black Bodies, White Bodies: Toward an Iconography of Female Sexuality in Late Nineteenth-Century Art, Medicine, and Literature." In *"Race," Writing, and Difference.* Ed. Henry Louis Gates, Jr. Chicago: University of Chicago Press, 1986.

Ginzburg, Carlo. "High and Low: The Theme of Forbidden Knowledge in the Sixteenth and Seventeenth Centuries." *Past and Present* 73 (1976): 28–41.

Goldberg, Jonathan. *James I and the Politics of Literature.* Baltimore: Johns Hopkins University Press, 1983.

——. *Writing Matter: From the Hands of the English Renaissance.* Stanford, Calif.: Stanford University Press, 1990.

Grafton, Anthony. *Defenders of the Text: The Traditions of Scholarship in an Age of Science, 1450–1800.* Cambridge, Mass.: Harvard University Press, 1991.

——. *Forgers and Critics: Creativity and Duplicity in Western Scholarship.* Princeton, N.J.: Princeton University Press, 1990.

——, ed. *Rome Reborn: The Vatican Library and Renaissance Culture.* Exhibition catalog. Washington, D.C.: Library of Congress, in association with the Biblioteca Apostolica Vaticana (Rome), 1993.

——, and Lisa Jardine. *From Humanism to the Humanities.* Cambridge, Mass.: Harvard University Press, 1986.

Greenblatt, Stephen. "Learning to Curse: Aspects of Linguistic Colonialism." In *Learning to Curse: Essays in Early Modern Culture.* New York: Routledge, 1990, 16–39.

——. *Marvelous Possessions: The Wonder of the New World.* Chicago: University of Chicago Press, 1991.

——. *Renaissance Self-Fashioning.* Chicago: University of Chicago Press, 1980.

Gross, Kenneth. "Satan and the Romantic Satan: A Notebook." In *Remembering Milton: Essays on the Texts and Traditions.* Ed. Mary Nyquist and Margaret W. Ferguson. New York: Methuen, 1987, 318–341.

Guillory, John. "Dalilah's House: *Samson Agonistes* and the Sexual Division of Labor." In *Rewriting the Renaissance: The Discourses of Sexual Difference in Early Modern Europe.* Ed. Margaret Ferguson, Maureen Quilligan, and Nancy Vickers. Chicago: University of Chicago Press, 1986, 106–122.

——. "From the Superfluous to the Supernumerary: Reading Gender into *Paradise Lost*." In *Soliciting Interpretation: Literary Theory and Seventeenth-Century Poetry.* Ed. Elizabeth Harvey and Katharine Eisaman Maus. Chicago: University of Chicago Press, 1990, 68–88.

——. *Poetic Authority: Spenser, Milton, and Literary History.* New York: Columbia University Press, 1983.

Hall, Joseph. *Another World and Yet the Same: Bishop Joseph Hall's* Mundus Alter et Idem. Trans. and ed. John Millar Wands. New Haven, Conn.: Yale University Press, 1981.

Hallyn, Fernand. *The Poetic Structure of the World: Copernicus and Kepler.* Trans. Donald M. Leslie. New York: Zone Books, 1990.

Hampton, Timothy. *Writing from History: The Rhetoric of Exemplarity in Renaissance Literature.* Ithaca, N.Y.: Cornell University Press, 1990.

Haraway, Donna. *Primate Visions: Gender, Race, and Nature in the World of Modern Science.* New York: Routledge, 1989.

———. *Simians, Cyborgs, and Women: The Reinvention of Nature.* New York: Routledge, 1991.

Harding, Sandra. "Feminism, Science, and the Anti-Enlightenment Critiques." In *Feminism/Postmodernism.* Ed. Linda Nicholson. New York: Routledge, 1990, 83–106.

———. *Whose Science? Whose Knowledge?* Ithaca, N.Y.: Cornell University Press, 1991.

Harriot, Thomas. *A Briefe and True Report of the New Found Land of Virginia.* 1590. Reprint with an intro. by Paul Hulton. New York: Dover, 1972.

Hawkes, Terence. "Swisser-Swatter: Making a Man of English Letters." In *Alternative Shakespeares.* Ed. John Drakakis. New York: Methuen, 1985.

Helgerson, Richard. *Forms of Nationhood: The Elizabethan Writing of England.* Chicago: University of Chicago Press, 1992.

———. *Self-Crowned Laureates: Spenser, Jonson, Milton, and the Literary System.* Berkeley: University of California Press, 1983.

Hodgen, Margaret. *Early Anthropology in the Sixteenth and Seventeenth Centuries.* Philadelphia: University of Pennsylvania Press, 1964.

Hodges, Devon. *Renaissance Anatomies of Fiction.* Amherst: University of Massachusetts Press, 1985.

Hooper-Greenhill, Eilean. *Museums and the Shaping of Knowledge.* New York: Routledge, 1992.

Horkheimer, Max, and Theodor W. Adorno. *Dialectic of Enlightenment.* Trans. John Cumming. New York: Continuum, 1972.

Hulse, Clark. "Spenser, Bacon, and the Myth of Power." In *The Historical Renaissance.* Ed. Heather Dubrow and Richard Strier. Chicago: University of Chicago Press, 1988, 315–346.

Hunter, Michael. *Science and Society in Restoration England.* Cambridge: Cambridge University Press, 1981.

Impey, Oliver, and Arthur MacGregor, eds. *The Origins of Museums: The Cabinet of Curiosities in Sixteenth- and Seventeenth-Century Europe.* Oxford: Clarendon Press, 1985.

Irigaray, Luce. "Is the Subject of Science Sexed?" Trans. Carol Mastrangelo Bové. In *Feminism and Science.* Ed. Nancy Tuana. Bloomington: Indiana University Press, 1989, 58–68.

Jacob, Margaret. *The Radical Enlightenment.* London: Allen and Unwin, 1981.

James I and VI. *The Political Works of James I.* Ed. C. H. McIlwain. Cambridge, Mass.: Harvard University Press, 1918.

Jameson, Fredric. "Of Islands and Trenches: Naturalization and the Production of Utopian Discourse." *Diacritics* 7 (1977): 2–21.

———. *Postmodernism, or, The Cultural Logic of Late Capitalism.* Durham, N.C.: Duke University Press, 1991.

———. *Signatures of the Visible.* New York: Routledge, 1992.

Jed, Stephanie. *Chaste Thinking: The Rape of Lucretia and the Birth of Humanism.* Bloomington: Indiana University Press, 1989.

Jones, Mark, with Paul Craddock and Nicholas Barker, eds. *Fake? The Art of Deception.* London: British Museum, 1990.

Jones, R. F. *Ancients and Moderns.* 1961. Reprint, New York: Dover, 1982.

Jordanova, Ludmilla. *Sexual Visions: Images of Gender in Science and Medicine Between the Eighteenth and the Twentieth Centuries.* Madison: University of Wisconsin Press, 1989.

Kahn, Victoria. "Habermas, Machiavelli, and the Humanist Critique of Ideology." *PMLA* 105 (1990): 464–476.

Karp, Ivan, and Steven D. Lavine, eds. *Exhibiting Cultures: The Poetics and Politics of Museum Display.* Washington, D.C.: Smithsonian Museum Press, 1991.

Kellein, Thomas. "How difficult are portraits? How difficult are people!" In *Cindy Sherman 1991.* Exhibition catalog. Basel: Kunsthalle, 1991.

Kelley, Donald R. *Renaissance Humanism.* (Twayne's Studies in Intellectual and Cultural History, no. 2.) Boston: Twayne/G. K. Hall, 1991.

Knapp, Jeffrey. "Eros as a Means of Empire in *The Faerie Queene* I." *English Literary History* 54 (1987): 801–834.

Kolodny, Annette. *The Lay of the Land.* Chapel Hill: University of North Carolina Press, 1975.

Kott, Jan. "*The Aeneid* and *The Tempest.*" In *The Bottom Translation.* Evanston, Ill.: Northwestern University Press, 1987.

Koyré, Alexandre. *Newtonian Studies.* Chicago: University of Chicago Press, 1965.

Kuhn, Thomas. *The Copernican Revolution: Planetary Astronomy in the Development of Western Thought.* Cambridge, Mass.: Harvard University Press, 1957.

———. *The Essential Tension: Selected Studies in Scientific Tradition and Change.* Chicago: University of Chicago Press, 1977.

———. *The Structure of Scientific Revolutions.* 2d ed. Chicago: University of Chicago Press, 1970.

Lakatos, Imre, and Alan Musgrave, eds. *Criticism and the Growth of Knowledge.* Cambridge: Cambridge University Press, 1970.

Lamming, George. *The Pleasures of Exile.* 1960. New York: Allison and Busby, 1984.

Langford, Jerome J. *Galileo, Science, and the Church.* 3d ed. Ann Arbor: University of Michigan Press, 1992.

Works Cited

Laqueur, Thomas. *Making Sex: The Body and Gender from the Greeks to Freud.* Cambridge, Mass.: Harvard University Press, 1990.

Lee, Rensselaer W. *Ut Pictura Poesis: The Humanistic Theory of Painting.* New York: Norton, 1967.

Leininger, Lorie Jerrell. "The Miranda Trap." In *The Woman's Part: Feminist Criticism of Shakespeare.* Ed. Carolyn Ruth Swift Lenz, Gayle Greene, and Carol Thomas Neely. Urbana: University of Illinois Press, 1980, 285–294.

Lévi-Strauss, Claude. *Tristes Tropiques.* Trans. John and Doreen Weightman. New York: Atheneum, 1981.

Lindberg, David C., and Robert S. Westman, eds. *Reappraisals of the Scientific Revolution.* Cambridge: Cambridge University Press, 1990.

Logan, Marie-Rose. "The Renaissance: Foucault's Lost Chance?" In *After Foucault: Humanistic Knowledge, Postmodern Challenges.* Ed. Jonathan Arac. New Brunswick, N.J.: Rutgers University Press, 1988, 97–109.

Long, Pamela O. "Humanism and Science." In *Renaissance Humanism: Foundations, Forms, and Legacy.* Ed. Albert Rabil, Jr. Philadelphia: University of Pennsylvania Press, 1988.

Loomba, Ania. *Gender, Race, Renaissance Drama.* Delhi: Oxford University Press, 1989.

Lyotard, Jean-François. *The Postmodern Condition: A Report on Knowledge.* Trans. Geoff Bennington and Brian Massumi. Minneapolis: University of Minnesota Press, 1984.

Maclean, Ian. *The Renaissance Notion of Woman: A Study in the Fortunes of Scholasticism and Medical Science in European Intellectual Life.* Cambridge: Cambridge University Press, 1980.

Marcus, George E., and James Clifford, eds. *Writing Culture: The Poetics and Politics of Ethnography.* Berkeley: University of California Press, 1986.

Marcus, George E., and Michael M. J. Fischer, eds. *Anthropology as Cultural Critique.* Chicago: University of Chicago Press, 1986.

Marin, Louis. *Utopics: Spatial Play.* Trans. Robert Vollrath. Atlantic Highlands, N.J.: Humanities Press, 1984.

Marjara, Harinder Singh. *Contemplation of Created Things: Science in* Paradise Lost. Toronto: University of Toronto Press, 1992.

Martz, Louis L. *The Poetry of Meditation: A Study in English Religious Literature of the Seventeenth Century.* 2d ed. New Haven, Conn.: Yale University Press, 1962.

McKeon, Michael. *The Origins of the English Novel.* Baltimore: Johns Hopkins University Press, 1987.

———. "Politics of Discourses and the Rise of the Aesthetic in Seventeenth-Century England." In *Politics of Discourse: The Literature and History of Seven-*

teenth-Century England. Ed. Kevin Sharpe and Steven N. Zwicker. Berkeley: University of California Press, 1987.

McLean, Antonia. *Humanism and the Rise of Science in Tudor England.* New York: Neale Watson Academic Publications, 1972.

Merchant, Carolyn. *The Death of Nature: Women, Ecology, and the Scientific Revolution.* San Francisco: Harper and Row, 1980.

Metzler, J., ed. *Sacrae Congregationis de Propaganda Fidei Memoria Rerum, Vol. I: 1622–1700.* Rome: Herder Press, 1972.

Mignolo, Walter D. "Misunderstanding and Colonization: The Reconfiguration of Memory and Space." *South Atlantic Quarterly* 92, 2 (1993): 209–260.

Milton, John. *Complete Poems and Major Prose.* Ed. Merritt Y. Hughes. New York: Macmillan, 1957.

Montalbano, William D. "Vatican Finds Galileo 'Not Guilty': Pope Admits Error in Rejecting Theory." *Washington Post,* 1 Nov. 1992, A40.

Montrose, Louis. "Celebration and Insinuation: Sir Philip Sidney and the Motives of Elizabethan Courtship." *Renaissance Drama* n.s. 8 (1977): 3–35.

———. "The Work of Gender in the Discourse of Discovery." *Representations* 33 (1991): 1–41.

Moss, Jean Dietz. "Galileo's *Letter to Christina:* Some Rhetorical Considerations." *Renaissance Quarterly* 36 (1983): 547–576.

———. *Novelties in the Heavens: Rhetoric and Science in the Copernican Controversy.* Chicago: University of Chicago Press, 1993.

Mowat, Barbara A. "Prospero, Agrippa, and Hocus Pocus." *English Literary Renaissance* 11, 3 (1981): 281–303.

Mukerji, Chandra. *From Graven Images: Patterns of Modern Materialism.* New York: Columbia University Press, 1983.

Mullaney, Steven. *The Place of the Stage: License, Play, and Power in Renaissance England.* Chicago: University of Chicago Press, 1988.

Mulvey, Laura. "A Phantasmagoria of the Female Body: The Work of Cindy Sherman." *New Left Review* 188 (July–Aug. 1991): 137–150.

Neill, Edward D. *English Colonization of America During the Seventeenth Century.* London: Strahan, 1871.

Newton, Isaac. *Principia.* Trans. Andrew Motte (1729), rev. Florian Cajori. Berkeley: University of California Press, 1962.

Nicolson, Marjorie Hope. *The Breaking of the Circle: Studies in the Effect of the 'New Science' on Seventeenth-Century Poetry.* New York: Columbia University Press, 1960.

———. *Science and Imagination.* Ithaca, N.Y.: Cornell University Press, 1956.

O'Malley, John. "The Discovery of America and Reform Thought at the Papal

Works Cited

Court in the Early Cinquecento." In *First Images of America: The Impact of the New World on the Old.* Ed. Fredi Chiapelli. Berkeley: University of California Press, 1976, 1: 185–200.

Orgel, Stephen. *The Illusion of Power: Political Theatre in the English Renaissance.* Berkeley: University of California Press, 1975.

———. "Making Greatness Familiar." In *The Power of Forms in the English Renaissance.* Ed. Stephen Greenblatt. Norman, Okla.: Pilgrim Books, 1982, 41–48.

———. "Prospero's Wife." *Representations* 8 (1984): 1–29.

Ormiston, Gayle, and Raphael Sassower. *Narrative Experiments: The Discursive Authority of Science and Technology.* Minneapolis: University of Minnesota Press, 1989.

Ovenell, R. F. *The Ashmolean Museum, 1683–1894.* Oxford: Clarendon Press, 1986.

Panofsky, Erwin. *Galileo as a Critic of the Arts.* The Hague: M. Nijhoff, 1954.

Parker, William Riley. *Milton: A Biography.* Oxford: Clarendon Press, 1968.

Parry, Graham. *The Golden Age Restor'd: The Culture of the Stuart Court.* Manchester: Manchester University Press, 1981.

Patterson, Annabel. *Censorship and Interpretation: The Conditions of Writing and Reading in Early Modern England.* Madison: University of Wisconsin Press, 1984.

Petrarca, Francesco. *Letters from Petrarch.* Trans. Morris Bishop. Bloomington: Indiana University Press, 1966.

———. *Petrarch's Letters to Familiar Authors.* Trans. and ed. Mario Cosenza. Chicago: University of Chicago Press, 1910.

Pietz, William. "The Problem of the Fetish, I." *Res* 9 (1985): 5–17.

———. "The Problem of the Fetish, II." *Res* 13 (1987): 23–42.

———. "The Problem of the Fetish, IIIa." *Res* 16 (1988): 105–123.

Prigogine, Ilya, and Isabelle Stengers. *Order Out of Chaos: Man's Dialogue with Nature.* New York: Harper and Row, 1984.

Purchas, Samuel. *Hakluytus Posthumus, or Purchas His Pilgrimes.* London, 1625.

———. *Purchas His Pilgrimage.* London, 1614.

Pye, Christopher. "The Sovereign, the Theater, and the Kingdome of Darknesse." *Representations* 8 (1984): 85–106.

Quilligan, Maureen. "Sidney and His Queen." In *The Historical Renaissance.* Ed. Heather Dubrow and Richard Strier. Chicago: University of Chicago Press, 1988, 171–196.

Rabil, Albert, Jr., ed. *Renaissance Humanism: Foundations, Forms, and Legacy.* Philadelphia: University of Pennsylvania Press, 1988.

Rajan, B. *Paradise Lost and the Seventeenth Century Reader.* Ann Arbor: University of Michigan Press, 1967.

Redondi, Pietro. *Galileo Heretic.* Trans. Raymond Rosenthal. Princeton, N.J.: Princeton University Press, 1987.

Reiss, Timothy J. *The Discourse of Modernism.* Ithaca, N.Y.: Cornell University Press, 1982.

———. *The Meaning of Literature.* Ithaca, N.Y.: Cornell University Press, 1992.

Rensburger, Boyce. "Optics: Galileo's 'Perfect' Lens of 1609." *Washington Post,* 13 July 1992, A2.

Riley, Denise. *"Am I That Name?" Feminism and the Category of "Women" in History.* Minneapolis: University of Minnesota Press, 1988.

Rose, Jacqueline. *"Hamlet*—the 'Mona Lisa' of Literature." In *Sexuality in the Field of Vision.* London: Verso, 1986, 123–140.

———. *"The Man Who Mistook His Wife for a Hat* or *A Wife Is Like an Umbrella*— Fantasies of the Modern and Postmodern." In *Universal Abandon? The Politics of Postmodernism.* Ed. Andrew Ross. Minneapolis: University of Minnesota Press, 1988, 237–250.

———. "Sexuality in the Reading of Shakespeare." In *Alternative Shakespeares.* Ed. John Drakakis. New York: Methuen, 1985, 95–118.

Rosen, Edward. *Copernicus and the Scientific Revolution.* Malabar, Fla.: Robert E. Krieger, 1984.

———, ed. and trans. *Three Copernican Treatises.* New York: Dover, 1958.

Rossi, Paolo. "Hermeticism, Rationality, and the Scientific Revolution." In *Reason, Experiment, and Mysticism in the Scientific Revolution.* Ed. M. L. Righini Bonelli and William R. Shea. New York: Science History, 1975, 247–273.

Salusbury, Thomas. *Mathematical Collections and Translations: The First Tome; in Two Parts.* London: William Leybourn, 1661.

Saunders, J. B. DeC. M., and Charles D. O'Malley, eds. *The Illustrations from the Works of Andreas Vesalius of Brussels.* New York: Dover, 1973.

Sawday, Jonathan. "The Flaying of Marsyas: Dissecting the Renaissance Body." In *Renaissance Bodies: The Human Figure in English Culture, c. 1540–1660.* Ed. Lucy Gent and Nigel Llewellyn. London: Reaktion Books, 1990, 111–135.

Schiesari, Juliana. "The Face of Domestication: Physiognomy, Gender Politics, and Humanism's Others." In *Women, "Race," and Writing in the Early Modern Period.* Ed. Margo Hendricks and Patricia Parker. New York: Routledge, 1994, 55–70.

Schwartz, Regina. "Rethinking Voyeurism and Patriarchy: The Case of *Paradise Lost." Representations* 34 (1991): 85–103.

Scott, Joan. "Gender: A Useful Category of Historical Analysis." In *Coming to Terms: Feminism, Theory, Politics.* Ed. Elizabeth Weed. New York: Routledge, 1989, 81–100.

Works Cited

Serres, Michel. *Hermes: Literature, Science, Philosophy.* Ed. Josue Harari and David F. Bell. Baltimore: Johns Hopkins University Press, 1982.

Shakespeare, William. *The Tempest.* Ed. Stephen Orgel. New York: Oxford University Press, 1987.

Shapiro, Barbara J. *Probability and Certainty in Seventeenth-Century England: A Study of the Relationships Between Natural Science, Religion, History, Law, and Literature.* Princeton, N.J.: Princeton University Press, 1983.

Shelton, Anthony Alan. "Cabinets of Transgression: Renaissance Collections and the Incorporation of the New World." In *The Cultures of Collecting.* Ed. John Eisner and Roger Cardinal. Cambridge, Mass.: Harvard University Press, 1994, 175–203.

Shumaker, Wayne P. *The Occult Sciences in the Renaissance.* Berkeley: University of California Press, 1975.

Sidney, Sir Philip. *A Defence of Poetry.* Ed. J. A. Van Dorsten. Oxford: Oxford University Press, 1966.

Snow, C. P. *The Two Cultures and A Second Look: An Expanded Version of the Two Cultures and the Scientific Revolution.* Cambridge: Cambridge University Press, 1964.

Solomon-Godeau, Abigail. "Suitable for Framing: The Critical Recasting of Cindy Sherman." *Parkett* 29 (1991): 112–115.

Spence, Jonathan. *The Memory Palace of Matteo Ricci.* New York: Penguin, 1984.

Spingarn, Joel. *Literary Criticism in the Renaissance.* New York: Columbia University Press, 1908.

Spivak, Gayatri. "Can the Subaltern Speak?" In *Marxism and the Interpretation of Cultures.* Ed. Lawrence Grossberg and Cary Nelson. Urbana: University of Illinois Press, 1988, 271–313.

Stafford, Barbara Maria. *Body Criticism: Imaging the Unseen in Enlightenment Art and Medicine.* Cambridge, Mass.: MIT Press, 1992.

Stevens, Henry. *Thomas Harriot and His Associates.* London, 1900.

Stewart, Susan. *On Longing: Narratives of the Miniature, the Gigantic, the Souvenir, the Collection.* Baltimore: Johns Hopkins University Press, 1984.

Strauss, Leo. *Persecution and the Art of Writing.* 2d ed. Chicago: University of Chicago Press, 1988.

Svendsen, Kester. *Milton and Science.* Cambridge, Mass.: Harvard University Press, 1956.

Tennenhouse, Leonard. *Power on Display: The Politics of Shakespeare's Genres.* New York: Methuen, 1986.

Todorov, Tzvetan. *The Conquest of America: The Question of the Other.* Trans. Richard Howard. New York: Harper and Row, 1984.

Trimpi, Wesley. *Muses of One Mind: The Literary Analysis of Experience and Its Continuity.* Princeton, N.J.: Princeton University Press, 1983.

Tuana, Nancy, ed. *Feminism and Science.* Bloomington: Indiana University Press, 1989.

Vaihinger, Hans. *The Philosophy of 'As-If.'* Trans. C. K. Ogden. New York: Harcourt Brace, 1924.

van der Straet, Jan (Stradanus). *"New Discoveries": The Sciences, Inventions and Discoveries of the Middle Ages and the Renaissance as Represented in 24 Engravings Issued in the Early 1580s by Stradanus.* Norwalk, Conn.: Burndy Library Publication no. 8, 1953.

———. *Nova Reperta.* Antwerp, ca. 1600.

Vickers, Brian, ed. *Occult and Scientific Mentalities in the Renaissance.* Cambridge: Cambridge University Press, 1984.

Wallace, William A., ed. *Reinterpreting Galileo.* (Studies in Philosophy and the History of Philosophy, vol. 15.) Washington, D.C.: Catholic University Press of America, 1986.

Webster, Charles. *The Great Instauration: Science, Medicine, and Reform, 1626–1660.* New York: Holmes and Meier, 1976.

Weinberg, Bernard. *A History of Literary Criticism in the Italian Renaissance.* 2 vols. Chicago: University of Chicago Press, 1961.

West, Cornel. "The Postmodern Crisis of the Black Intellectuals." In *Cultural Studies.* Ed. Lawrence Grossberg, Cary Nelson, and Paula Treichler. New York: Routledge, 1992, 689–705.

Westfall, Robert S. *Essays on the Trial of Galileo (Studi Galileiani* no. 5). Vatican City: Vatican Observatory Publications, 1989.

Westman, Robert S. "Proof, Poetics, and Patronage: Copernicus's Preface to *De Revolutionibus.*" In *Reappraisals of the Scientific Revolution.* Ed. David C. Lindberg and Robert S. Westman. Cambridge: Cambridge University Press, 1990, 167–205.

White, Hayden. "The Forms of Wildness: Archaeology of an Idea." In *Tropics of Discourse: Essays in Cultural Criticism.* Baltimore: Johns Hopkins University Press, 1978, 150–182.

Williams, Linda. *Hardcore: Power, Pleasure, and the Frenzy of the "Visible."* Berkeley: University of California Press, 1989.

Williams, Raymond. *The Country and the City.* New York: Oxford University Press, 1973.

———. *Keywords.* 2d ed. New York: Oxford University Press, 1983.

Wilson, Luke. "William Harvey's *Prelectiones:* The Performance of the Body in the Renaissance Theatre of Anatomy." *Representations* 17 (1987): 62–95.

Wind, Edgar. *Pagan Mysteries in the Renaissance.* 2d ed. New York: Norton, 1968.

Wright, A. D. *The Counter-Reformation: Catholic Europe and the Non-Christian World.* New York: St. Martin's Press, 1982.

Yates, Frances A. *Giordano Bruno and the Hermetic Tradition.* Chicago: University of Chicago Press, 1964.

Zamora, Margareta. "Abreast of Columbus: Gender and Discovery." *Cultural Critique* 18 (1991), 127–149.

Index

Index

About the Author

Denise Albanese is Associate Professor of English and Cultural Studies
at George Mason University.

Library of Congress Cataloging-in-Publication Data

Albanese, Denise.
New science, new world / by Denise Albanese.
p.   cm.
Includes bibliographical references and index.
ISBN 0-8223-1759-1 (cloth : alk. paper). — ISBN 0-8223-1768-0 (pbk. : alk. paper)
1. English literature—Early modern, 1500–1700—History and criticism.
2. Literature and science—England—History—17th century.   3. Galilei, Galileo,
1564–1642.   Dialogo . . . dove . . . si discorre sopra i due massimi sistemi del
mondo.   4. America—Discovery and exploration—Historiography.   5. Shakespeare,
William, 1564–1616.   Tempest.   6. Bacon, Francis, 1561–1626.   New
Atlantis.   7. Donne, John, 1572–1631.   Conclave ignati.   8. Milton John, 1608–
1674.   Paradise lost.   9. Science—History—17th century.   10. Imperialism in
literature.   11. Science in literature.   I. Title.
PR438.S35A43   1996
820.9′356—dc20   95-47757   CIP